Keeping Live Corals and Invertebrates

Also by ROBERT P. L. Straughan:

Keeping the Dwarf Sea Horse
Keeping Sea Horses
Adventures in Marine Collecting
The Marine Collector's Guide
Exploring the Reef
Jet Safari to Africa
Build a Jungle Zoo
Adventure in Belize
The Salt-Water Aquarium in the Home

Keeping Live Corals and Invertebrates

Robert P. L. Straughan

South Brunswick and New York: A. S. Barnes and Company
London: Thomas Yoseloff Ltd

A. S. Barnes and Co., Inc.
Cranbury, New Jersey 08512

Thomas Yoseloff Ltd
108 New Bond Street
London W1Y OQX, England

Library of Congress Cataloging in Publication Data

Straughan, Robert P L
Keeping live corals and invertebrates.

Bibliography: p.
Includes index.
1. Marine aquariums. 2. Marine invertebrates as pets. I. Title.
SF457.1.S75 639'.34 74-9311
ISBN 0-498-01635-8

All photos are by the author unless otherwise designated.

PRINTED IN THE UNITED STATES OF AMERICA

Contents

Preface

EACH YEAR WE LEARN A LITTLE MORE ABOUT THE SEA. MAN HAS EXPLORED the ocean to its greatest depths. He has built houses under the sea to study its inhabitants, but this is costly and time consuming when compared to what can be learned about the sea by bringing a portion of it into the laboratory or classroom or into our homes.

The aquarium can be as great a tool for learning as the underwater habitat. Place a piece of live coral in a glass aquarium and one can study it intimately with a magnifying glass or low-power microscope on a round-the-clock basis if desired. Bring a chunk of the sea floor alive into the laboratory and the scientist or student can learn much more about it than could ever be learned from observing it under the sea. In the comfort of the laboratory or home study, the light can be controlled exactly. There are no currents or turbid waters to distort the view. In the quiet of the aquarium, strange and sometimes minute organisms will suddenly appear from within a live rock or coral. Eggs and larvae will hatch out and grow into recognizable forms. Many of these would have been entirely missed by the underwater observer. Not that the underwater observer is unnecessary, for his work is vital to understanding the sea, but rather, we must combine the marine aquarium with research and realize its enormous value.

It is more practical to bring a portion of the sea floor or living reef into the classroom where dozens or hundreds of students can observe and study it at close hand than to bring the same hundreds of students to the sea floor. We have now made sufficient progress in keeping corals and invertebrates alive, even for the inland aquarist, that nearly anyone can keep a good portion of it alive for a reasonable length of time. This book tells how it can be done, what to expect in longevity of a transplanted reef, and what the future holds in keeping and maintaining the wonders of the sea—even from the great depths.

Robert P.L. Straughan

Acknowledgments

I WISH TO EXPRESS MY APPRECIATION TO MR. THOMAS YOSELOFF, JULIEN Yoseloff, Warren Prince, and all those who contributed with information or worked on the production of this book.

Keeping Live Corals and Invertebrates

1

A Reef in the Living Room

THE FASCINATING SALTWATER AQUARIUM HOBBY HAS NOW PROGRESSED TO the point that it is possible to keep a small segment of a living coral reef, complete with fishes, right in the living room. You can keep a portion of an entire coral reef (in miniature of course) and keep it in perfect condition for an indefinite period of time governed only by

The coral reef, an oasis under the sea, is a complex structure. Yet with modern technology, even a scene such as this could be duplicated in the aquarium. It would be a simple matter to collect the school of grunts, but keeping the corals alive would require expert knowledge and a source of clean sea water.

your expertise as an aquarist. All the fascination of the coral reef, the interesting corals, the swaying gorgonians, the sea anemones, sea cucumbers, sea urchins, and even the gaudy fishes can be transplanted to the aquarium and kept alive for periods of several months to several years or longer.

Of course we don't know all there is to know about keeping corals, fish, or anything from the sea, and we *never* will know everything about it, for the ocean is a complex medium and the coral reef is even more complex with its intricate balances and symbiotic relationship with the sea, without which it could not survive. But we do know enough to keep both the ocean and the reef in the confines of an aquarium and keep our "miniature ocean" in good shape, for a reasonable length of time.

Recent advances in the aquarium hobby, such as high-quality synthetic sea salts that allow the hobbyist to make his own sea water and the modern all-glass aquarium which was first invented by myself, have helped the saltwater aquarium hobby take a firm hold throughout the world. In addition, there has been a gradual increase in knowledge about keeping marine creatures in captivity due largely to the fact that there are now hundreds of thousands of marine aquarists through-

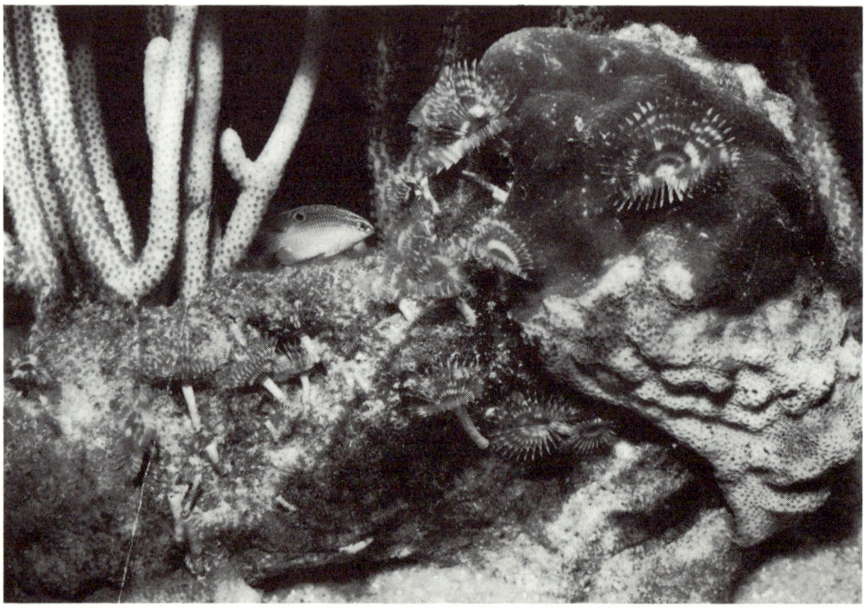

Much less complex is this singular piece of coral rock housing a colony of interesting tube worms. A small patch of live coral is growing at the top. Live gorgonians in the background enhance the scene. A small segment of the reef such as this can easily be kept by the aquarist.

Undersea photo? No! This is actually a scene from a fifteen-gallon aquarium I set up with live plants, corals, and alcyonarians.

out the world compared to just twenty or tweny-five years ago when there were probably no more than five hundred saltwater aquarists in the entire world! If one were to count the hundreds of thousands, perhaps millions, of aquarists who keep sea horses, the thousands of students keeping marine organisms alive in the classroom, then the ranks of the saltwater hobbyists would easily swell into the millions and perhaps this figure would be more accurate.

Consequently a great deal of knowledge and technical know-how have been amassed. Even the breeding and rearing of saltwater fish has been achieved. The saltwater aquarist has graduated from the dead, nearly sterile aquarium with bleached corals and dead seashells to the live coral aquarium, or at least he will have graduated by the time he finishes this book, for a live coral aquarium is so much more interesting and rewarding than a sterile tank that there is hardly any comparison. Keeping corals alive opens up a vast new world in the aquarium field, for now we can enjoy the ultimate in saltwater, the keeping of a living coral reef in our home.

The old-style aquarium of bleached, snow-white corals and pale fish

The aquariums of snow-white, bleached corals are beautiful, but they are slowly being replaced by the live coral aquariums when live corals can be obtained.

will someday be passé, for once the aquarist has tasted the delights of a live coral tank with its wondrous living polyps that move and change right before your eyes, the secretive brittle starfish that emerge from beneath the coral to feed, the multitudes of small living seashells that live on and within the corals, the tiny crabs, the sea squirts, and all the wonderful little creatures that make up the coral reef community, he will find that his whole aquarium is alive, not just the fish, as is the case with the bleached-coral aquarium.

The coral reef is a pulsating, living thing from the very outside of the coral heads, which contain the living polyps, to the very core of the coral itself. The inside of the coral is riddled with tube worms, boring clams, mollusks, barnacles, boring worms, and numerous crabs and shrimps. Even coral that has died is still pulsating with life of all types. The coral polyps may all be dead and the coral itself will cease to grow, but its hard skeleton of calcium carbonate will provide a home for multitudes of sea creatures until it finally crumbles to the sea floor. But this is merely an aging process of the sea. The coral lives, then

A giant coral head twenty feet across is a living thing, not only on its outer surface, which contains the coral polyps, but throughout its interior which hosts a wide variety of marine life. Clams, mollusks, shrimp, crabs, sponges, barnacles, and numerous small fish live inside the coral, which is often hollow or riddled with large holes and crevices.

The ultimate in a live coral aquarium is to keep a full complement of fish as well. It can easily be done if the fish are disease free. Two beautiful royal grammas *(Gramma loretto)* cavort over a live coral head in my aquarium.

dies and settles to the sea bottom where more corals grow upon it to continue the cycle. In this book, I will treat the keeping of live "rocks" as well as corals, for both are an intricate part of the coral reef, the latter being merely a part of the living coral reef that has died.

The sterile tank that I have recommended in other books will still be useful and serve its purpose for those who cannot obtain live corals, but eventually in future years, it may become obsolete, for I feel that the live coral tank is the tank of the future. It is far superior in interest and beauty to the dead-coral tank. In addition, the fish will sparkle. They will even spawn, for they will feel right at home. Fish that were formerly considered difficult to keep will flourish in the live coral tank, for the live corals and rocks will furnish them with algaes and nutrients that will sustain them, keeping them in perfect health.

You will find the live coral tank so fascinating that you may sit in front of it for hours watching the drama of undersea life unfold before your very eyes. You will see things you never knew were even in the

A forest of gorgonians on a shallow reef near Placentia, Belize. It would be possible to keep most or all of them alive in the saltwater aquarium with proper aeration, filtration, and feeding, but the thicker specimens at lower left are preferred, because they are quite hardy. A colonial tunicate is in the center.

An unwanted visitor is the ubiquitous barracuda, barely visible at the top center of this pretty reef scene. Barracuda often stay just out of sight, and the novice may never see one at all.

Closer view reveals the delicate beauty of the bleached corals. But alive, they would be far more interesting and they might even grow. Sea fan has been cleaned and cured for the aquarium by removing the outer coating.

aquarium, for one of the bonuses of the live coral tank is that there are all kinds of tiny sea creatures living within the coral. Some may hatch out after a few months, others will grow from tiny babies too small for you to see at first, but eventually you will spot them moving about the tank. You may see tiny crabs or shrimp, little sea anemones, or even baby jellyfish, baby fish, or seashells that will hatch out in the aquarium. The living coral aquarium will never cease to amaze you. You will find it far more interesting than television!

Furthermore, it is not difficult to keep or maintain if you will follow the simple procedures outlined in this book. Not only will you be able to keep the live corals, live rocks, seashells, and all kinds of invertebrates, but you can also keep a full complement of exotic coral fishes, those colorful, exquisite living gems of the sea, making the coral reef complete, for what is a coral reef without its fishes? The fishes will be more contented than they ever were in the sterile tank, and if all goes well, you should have more success with your marine aquarium than you ever thought possible in the past.

2
Setting Up the Reef

THE LIVE CORAL AQUARIUM IS SET UP WITH THE STANDARD UNDERGRAVEL filter system, which I first devised in 1952. An outside filter with activated carbon is optional for a small tank but recommended for any tank larger than twenty gallons. The carbon will keep the water

A modern all-glass aquarium I designed, with sub-sand filter in place. The sub-sand filter should fill the bottom of the tank as closely as possible for best results. However, even a half panel can be used if set up properly. Note the wide frame at the top of the tank. This is preferred as it hides the water line.

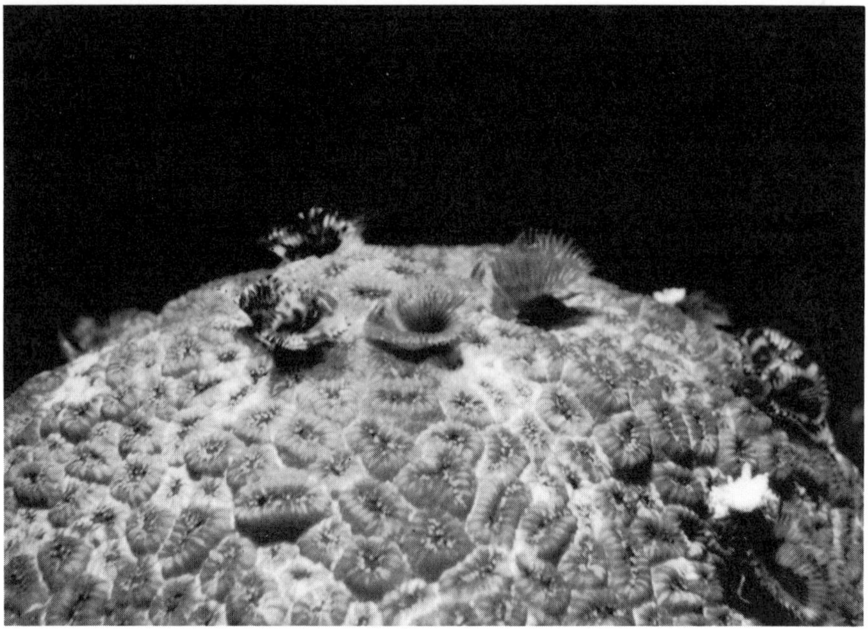

A live coral head with tube worms is much more interesting in the aquarium than a piece of dead, bleached coral.

sparkling white and clean, which is how it is on the reef. Take away the carbon and your tank may turn dingy brown or yellow, hardly a pleasing picture, unless you change a third of your water each month. Here is the simple procedure for setting up your aquarium.

Equipment Needed

1. A dependable air pump; a good vibrator type will do.
2. Air line tubing and a plastic four-way gang valve.
3. An all-glass aquarium of ten- to twenty gallon size. An aquarium stand.
4. An undergravel filter to fit the aquarium.
5. Tight-fitting cover glass.
6. A fluorescent aquarium light.
7. An aquarium heater.
8. A thermometer and hydrometer (or combination).
9. Aquarium sand, silica sand, or regular aquarium gravel, natural shell sand, coquina sand, coral sand, or a mixture. Figure about two pounds per gallon.
10. An outside filter, power filter, air driven filter, or an inside bottom filter. Some activated carbon, a pound or two.

11. A collection of live rocks and corals, invertebrates, and fish. Don't get these until the tank is all set up for two weeks!
12. This book, and a copy of my other book, *The Salt-Water Aquarium in the Home* (covers all aspects of the saltwater hobby).

Now the work begins. Place a piece of half-inch or three-quarter-inch *exterior* plywood on your aquarium stand and set the tank on top of it. It is best to paint the plywood to preserve it. Also it will look better. Place the aquarium in a draft-free location, preferably near a window where it will receive some sunlight at some time of the day, if at all possible. Live corals contain zooxanthellae and algae within their polyps and the algae require sunlight to carry on the bodily processes of the corals. If a location cannot be obtained with sunlight, then set the tank in a well-lighted area and use artificial light.

Next, fill the aqaurium with fresh water to check for leaks. Let it stand at least overnight. Then siphon out the water or dip it out with a gallon plastic measuring scoop. Never tip the aquarium to get the water out, since this may make it leak. Rinse the aquarium while you are emptying it with a new, clean sponge that has been carefully rinsed and that has never been used around the house for cleaning. Soap or household oils can be deadly to your aquarium, so use a new sponge to be certain. You must also be certain that your hands contain no oil or soap. To be safe, you should rinse them thoroughly up to the elbows to remove all traces of soap. Do this before placing hands in the tank, even when just rinsing the tank out.

When the tank is drained dry, use the sponge to wipe up any excess water. Then carefully wipe all the inside surface of the tank to remove any scum from the glass. Also wipe the rubber sealer on the seams of the tank to remove any film there.

If a stainless-steel tank is to be used, the tank should be thoroughly dried and the seams tightly sealed with silicone rubber or silastic aquarium cement. This should be allowed to dry for a few days and the tank should then be soaked to check for leakage. A cupful or two of salt should be added to cure the seams. The tank should then be emptied and wiped clean. But if at all possible, an all-glass tank should be used. They are very common today and easily available, or you can make one by simply gluing the glass together, following instructions supplied with the silicone rubber.

After the tank has been soaked and rinsed, wipe the inside dry with paper towels or lint-free cloth. This will help to remove any oil or scum from the glass that you may have missed. Then set the sub-sand filters in place. Usually there will be two of them. Push them tightly into the corner with the filter stem in back. Make certain that the filter stem is firmly in place. The air line tubing should be attached

Next step is to cover the undergravel filter with well-washed sand, to a depth of about two inches. The bubbler stems are attached to a gang valve that goes to the air pump. The pump should be placed above the aquarium to avoid water siphoning into it during power failure. The water may be added next by placing a sheet of Saran Wrap on the sand, holding it in place with small rocks. This will keep sand from being disturbed. This tank has a wooden frame coated with epoxy paint.

to the filter stems and fastened to the gang valve, which is hung on the back or side of the tank.

Adding the Sand

There is a great deal of controversy about which type of sand is best for use with an undergravel filter. In the past, I have recommended silica sand with grains about 1/16 to 1/32 inches long, or just coarse enough so that it will not pass through the slots in the sub-sand filter. The proper size of the sand rather than the type of sand is the most critical factor. If the sand is too fine, it will sift through the sub-sand filter, clogging the air space beneath it and greatly reducing or destroying its filtering action. If the sand is too coarse, say pebble size or larger, you won't get any filtration at all, for the water will simply pass through both the sand and the filter and the dirt will settle down to the bottom of the tank and cause severe problems. Therefore, you must use the right size sand. Ordinary aquarium gravel of the type used in fresh-water aquariums is usually alright if it does not contain any minerals. I have always recommended silica sand because it is rather pure and very inert in saltwater. But I also recommend coquina shell sand, crushed coral sand, or natural beach sand (a little on the coarse side) if these are taken from a clean, unpolluted area. Some aquarist use dolomite or calcite gravel and these may be alright, but I would certainly recommend washing and soaking them for two weeks in saltwater to leech out any toxic products from them.

All sand to be used should be washed carefully by placing it in a clean bucket that has never contained anything but water. Place about five inches of sand in the bucket. Then fill with water from the garden hose or faucet. Reach your hand down to the bottom and swirl the sand around thoroughly. Then, before it has settled, pour off the fine sand and dirty water until you come to the main portion of sand. Do this about five times or until the water remains perfectly clear when you fill the bucket and stir the sand. The sand is then ready for use. If you suspect that the sand may contain impurities, after you have washed it throughly, fill the bucket with saltwater and let it stand for a couple of days. Then rinse again and pour off all the water. The sand may now be added to the aquarium. (Sand should be washed outside to avoid clogging drains)

Hold the filter plate down tightly against the corner and place the sand on top of it, smoothing it out so that it is slightly deeper toward the back. Put it about two inches deep, making a few slight mounds so the bottom will appear more interesting.

Adding the Water

You are now ready to add the water. Place a large, well-rinsed saucer or plate on the sand and pour the water gently into this, letting it flow gently onto the sand. This will keep the water from making a hole in the sand or washing down into the sand and undergravel filter. Fill the aquarium about one and one-half inches from the top or to the top frame line if your tank has a frame. By filling it to the bottom of this top frame, you will automatically have a salinity guide to go by so that you may keep a check on water that has evaporated. When the water gets down below the frame, you will know that some of it has evaporated so you can add fresh water to replace it. (This will be dealt with later.)

Fresh sea water is best of course and if this can be obtained, it should be used. But if it can't, you can use a good brand of artificial sea water. The water should have a specific gravity of 1.025 or slightly less. Many aquarists keep it at 1.023 or 1.024 depending upon the accuracy of the hydrometer, but at any rate, it should not be higher than 1.025 or lower than 1.020 for best results. If you are using artificial sea water, you can mix this in a clean bucket that has never contained soap, chemicals, or other products. Place a few pounds of salt in the bucket and fill it with water. Reach to the bottom and stir rapidly in a circular motion, until the salt has dissolved. Then let it settle and check it with a hydrometer, adding more salt or more water until you get the proper reading. Let the water settle out crystal clear before adding it to the aquarium.

Or if you like, you can mix the artificial sea water right in the aquarium itself. In this case, the filters and sand are not set in place until the water is mixed to the proper salinity and has cleared. The aquarium should be filled to about four inches from the top to allow for the sand. When the water has been mixed and is clear, you may now add the undergravel (sub-sand) filters. Attach the air tubing to the bubbler stem of the filter, then push the filter to the bottom of

If you collect your own specimens and do not have enough sea water to fill the aquarium, fill the aquarium part way as shown nere. Then, mix up some artificial sea water. Let it age for a week, and add a small amount to the tank each day until the aquarium is full. This will lessen the transition from the natural to the artificial.

the tank, shaking it slightly to remove any air bubbles that may lodge on it. Then, holding the filter tightly on the bottom and up against the corner, place a saucer full of washed sand on top of it. Do the same to the other panel if there are two panels to the filter. The balance of the sand may then be evenly distributed on top of the filters.

The air pump should then be turned on and the valve regulated so that the bubbler stems of the undergravel filter will give off a strong surge of bubbles. This will both aerate and filter the aquarium, for as water is forced up the bubbler stem tube, it creates a movement of water under the sand, drawing water down into the sand from the aquarium. This traps fine bits of dirt and debris in the sand where they are held to eventually decompose by bacterial action. Any gases that form are immediately dispelled by the bubbler stem. Without the undergravel or sub-sand filter, deadly hydrogen sulfide gas would soon form at the bottom of the tank if any sand is used. This would be catastrophic to live coral or any sea creatures. I was the first in the world to promote the use of the sub-sand filter in salt water aquariums. At first, there was considerable controversy over its use, but it has become thoroughly accepted in both salt- and fresh water and there are millions of them in use. Reams have been written about how and why it works describing the bacterial action of the filter, some supporting my views, others against them. Since I was the first to use it for saltwater and since I have set up thousands of saltwater aquariums with it over the past twenty years, I believe in my own findings rather than some of the so-called experts with their quasiscientific theories. You will hear all kinds of things about which sand to use, which salt to use, nitrates and nitrites, ozonizers, protein skimmers, etc., and you will find that I don't recommend too many of them, having tried them all. I don't recommend oyster shells because they usually come from the most polluted of waters. I am not too keen on dolomite or calcite because I feel that they dissolve too rapidly and make the water excessively hard. I don't recommend ozonizers because I haven't seen them do much.

You will also hear a great deal about inoculating your aquarium with some bacteria or water from an established tank. I am not enthused about this either. I never inoculate natural sea water tanks, for they have all the live bacteria they need, and I caution aquarists about taking some sea water or sand from another aquarium to "start" the bacteria in their own tank. With artificial sea water, the tank is totally sterile when it is first set up. Perhaps this is what gave rise to the idea of seeding it with germs. A few gallons of fresh sea water will give it all the healthful bacteria it needs. This should be added after the tank has been filtering for a few days. It will give life to the sterile

I do all of my own private research and have maintained hundreds of salt-water aquariums. This battery of aquariums was just one of several at my experimental shop in Central Florida near Ocala.

Part of my lab contains delicate balance scales (triple beam), various drugs and chemicals, microscopes, and formalin for preserving specimens. The low power stereo microscope proved the most useful.

water so that when you are ready to add your corals the tank will be in a healthful condition. But don't take for granted that water from any other aquarium will do. Many aquariums are infested with lethal parasites or dinoflagellates and if you add these to your tank you will be starting a culture of potentially harmful organisms that may be most difficult to eradicate. Parasites and disease are a real problem in saltwater aquariums, and worldwide pollution as well as the fact that fish and invertebrates from the oceans of the world are often mixed together make the disease factor even more complex. For this reason, I stress that you secure clean sea water from an unpolluted area to fill your aquarium or if you are using artificial water, obtain your sea water from your dealer requesting some that has had no fish in it if possible.

Carbon Filter

When your tank is all set up, the addition of a small inside filter containing well-washed activated carbon will keep your water sparkling clean and white. Or an outside filter such as an Eheim power filter or dynaflo filter with carbon and nylon floss may be used. Both of these

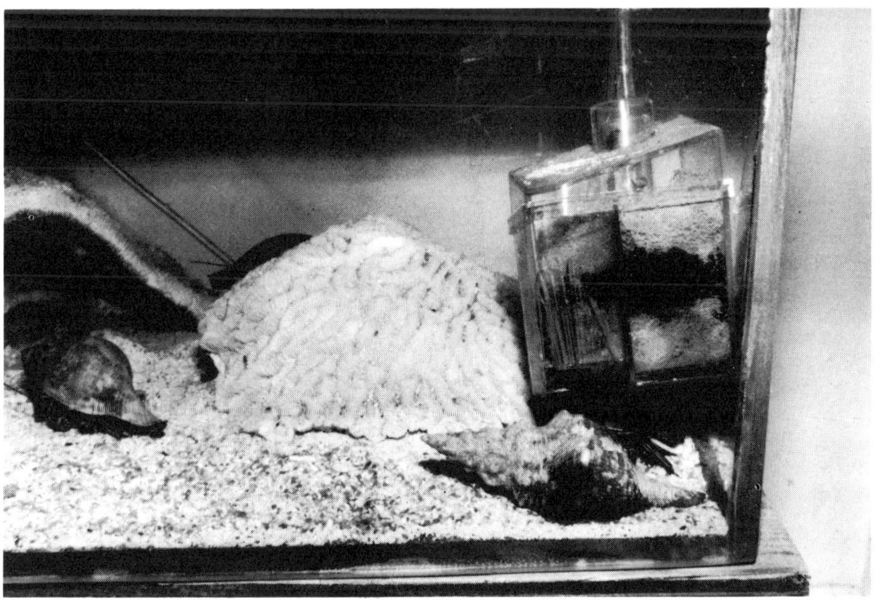

Activated carbon will help keep the water clear. It may be placed in an outside filter on the back of the aquarium or in a small inside bottom filter such as the one shown above. The filter is placed in the back corner behind a rock.

Sea fans, sponges, and other material that would be difficult to keep alive even in the invertebrate tank can be dried and painted their natural color then placed in a shadow box in back of the tank.

The intricate webbing of the sea fan represents some of nature's finest handiwork, but sea fans are difficult to keep alive in the aquarium at present. If you wish to try a live specimen, select a small six-inch fan attached to a rock. Otherwise, it would be best to place the sea fans in a shadow box behind the aquarium.

are excellent and they will assure good circulation of the water as well as superb filtration. The siphon stem from the outside filter should reach all the way to the bottom so that it is just above the sand. It should have a screen or slots at the bottom to avoid sucking up small fish into the filter. It should also be anchored firmly in place so that it won't move and possibly lose its suction. The Eheim has suction cups for this purpose, which is an excellent idea because it allows you to anchor the siphon tube and the outlet in any position desired without worry of the tubes getting out of place. But if there are no suction disks on your filter, you can position the tube or tubes in place and then secure them against the back of the tank with a well-cleaned rock or coral. Activated carbon is one of the finest filtering mediums in the world. It removes carbon dioxide from the water as well as other gases and impurities and will get the water so clear that it will be practically invisible. It should be well washed and aged in saltwater for a couple of weeks if possible, before use.

A small tank can operate fine with a simple inside bottom filter of

A magnetic-drive, motor-driven filter is excellent for the larger tanks, because it gives good circulation of the water. Nylon wool is placed on the bottom and is covered with well-aged activated carbon. The siphon stem reaches all the way to the bottom, which is best for efficient filtration.

The Eheim Power Filter is recommended for large aquariums and can be obtained in any size, operating tanks up to five hundred or even a thousand gallons. It furnishes perfect aeration and filtration. It is distributed in the United States by Kordon Corporation, Hayward, California.

carbon. The bubble-up type by Metaframe with the grooved slots at the bottom edge is best because it picks up dirt from the bottom of the filter rather than from the top and this is a decided advantage. The carbon filter will also aerate as well as filter the tank. Regardless of which kind of carbon you use and regardless of claims of the label, I would recommend that you wash and age all carbon to be on the safe side. Fresh carbon can cause problems, making too rapid a change in the water, which is why I suggest that it be well washed and aged.

Now that the aquarium is set up, you can put on the cover glass and place the aquarium light on top of this. The light may be turned on to check the clarity of the water. The tank should be allowed to filter and aerate for a week if natural sea water is used or at least two weeks if artificial water is used. If you live in a northern climate and winter is approaching, an aquarium heater should be inserted and set for approximately seventy-five degrees F. Your aquarium is now ready for its inhabitants.

3

Stocking the Aquarium

YOU ARE NOW READY TO TRANSFORM YOUR AQUARIUM INTO A LIVING coral reef, or a reasonable facsimile. If you live within driving distance of the coral reefs of Florida or live near the ocean in the Caribbean or anyplace in the world near the reefs, it will be simply a matter of a collecting trip or several to obtain the desired specimens. You will need face mask, flippers, snorkle, gloves, a floating live well, buckets, and several Styrofoam fish boxes with plastic bags. You will also need a boat unless the reefs are very close to shore in the area in which you will collect your specimens. (See chapter 5 for complete collecting instructions, if you plan to collect your own specimens.)

If you do not live within driving distance of the ocean, you will then have to buy your live corals and specimens either from your local aquarium dealer or have them shipped to you direct from the collector. Either way is alright and both have advantages and disadvantages. There is a lot in favor of getting your specimens from your local dealer for you can see the specimens before you buy them, picking out only those which are in near-perfect condition. You could select a half-dozen grapefruit sized or smaller, live corals, several anemones, an assortment of small crabs and snails, and several live rocks, with a few live gorgonians if they are obtainable. This will give you an exciting aquarium and a good segment of a coral reef in miniature. Do not buy any fish at present.

The live rocks differ from the live coral in that they contain no coral polyps. They are usually dead corals but they are alive with a huge variety of marine life including mollusks, sea squirts, algaes, barnacles, and plume worms, as well as a host of smaller organisms. They are as much a part of the coral reef as the corals themselves and equally fascinating to keep in the aquarium. If possible, obtain some

The magnificent marine jewel fish *(Microspathodon chrysurus)* lives in an aquarium of live finger coral and algae. Round ball on rock at right is *Valonia ventricosa*, the largest single plant cell in the world.

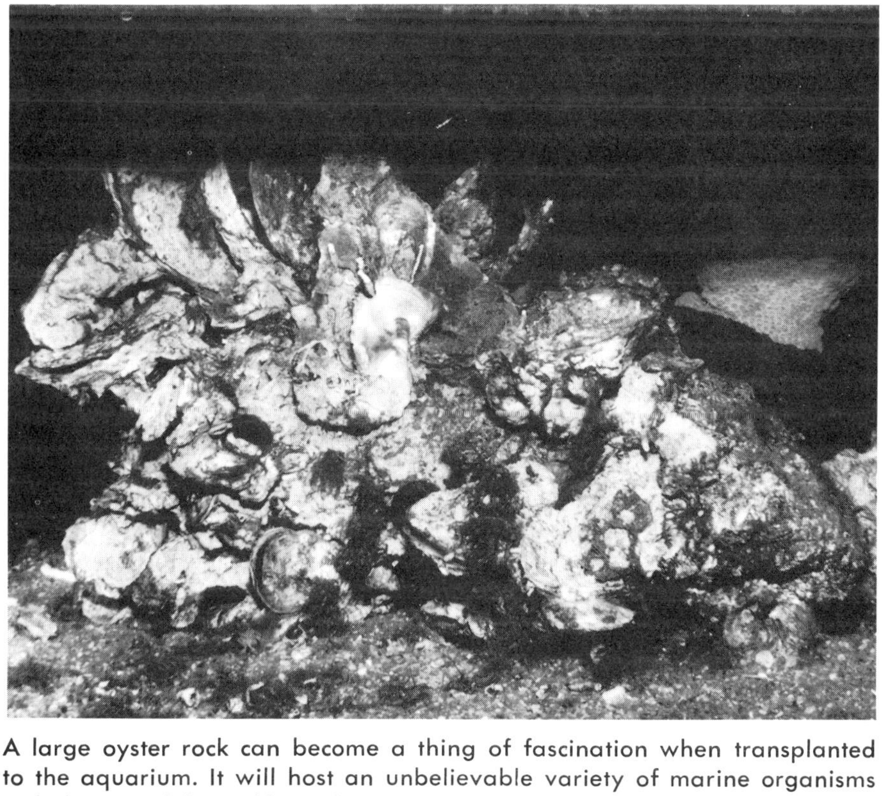

A large oyster rock can become a thing of fascination when transplanted to the aquarium. It will host an unbelievable variety of marine organisms including starfish, nudibranchs, sea squirts, and perhaps little blennies.

with numerous tube worms or plume worms as they are popularly called. These are most interesting and a real delight to watch. They will disappear right before your eyes, retracting instantly at the slightest disturbance, and there are a tremendous variety of them. Some even give off a smoke screen when disturbed and look like a tiny volcano under the sea.

If your local dealer doesn't handle live corals or cannot obtain them for you, then you will have to import your own from a marine collector. You can find their names and addresses in one of the aquarium magazines or perhaps your dealer will supply you with a reliable shipper. Some collectors do not ship to hobbyists because they sell only wholesale, but some do ship to individuals, and these should be contacted for prices and shipping requirements. The specimens will most likely be shipped by air freight to your nearest large commercial airport and you will be expected to pick up the specimens at the airport and pay the freight. On small orders, you may be able to have them shipped air mail, special delivery, direct to your home. Or if your collector isn't too far away he might be able to ship them to you by Greyhound bus. There will usually be a minimum order of perhaps thirty dollars to fifty dollars for each shipment or more so that if this is too much for you, then you may be able to split the cost with another person who wants to try a live coral tank. If you order live corals from a collector, you could also order a few live fish so you can get them all at the same time with one shipment. Request that the shipper send your fish and specimens in natural sea water rather than artificial and tell him to send plenty of water with the shipment. You will need this to give your aquarium the necessary minerals and plankton as well as healthful bacteria. The freight charges should be anywhere from twelve dollars to twenty dollars more or less.

This book will tell which corals do best and will list a number of invertebrates to choose from so you will have some idea about what to order. But you may have to take what you can get depending on how many collectors are supplying specimens at the time you need them. Pacific marine fish shippers number in the thousands and many are now shipping both live rocks and corals, so if you can't get Atlantic specimens then you can try the Pacific types. All are fascinating, regardless of from which ocean they come, for we have barely scratched the surface in keeping marine invertebrates. Some are so strange they look like something from another planet, but even the most common ones like the starfish, barnacle, or sea urchin are marvelous to observe in the aquarium, especially to the novice.

One thing is for certain, whether you collect your specimens, buy them locally, or have them shipped to you, be certain to have your

A simple invertebrate aquarium containing several live "rocks." The live rock is simply a piece of dead coral that has become overgrown with algaes, tube worms, and mollusks.

Tiny live corals growing on a piece of dead coral rock. Sea anemones add to the interest of the rock.

aquarium all set up and in operation for at least a week or two before you get any live specimens. Never go out and get the specimens and then come home to set up the tank. You will probably lose all or a good portion of the specimens unless you have had a great deal of experience. Get the tank ready first.

Adding Specimens to the Aquarium

Assuming you were successful in obtaining an assortment of corals and live rocks, your next step is to place them in the aquarium. Never dump them into the tank, for each rock and coral must be carefully handled to avoid damage to it. When you first arrive home with your specimens, set the shipping box or carrying box near the aquarium and insert an air stone or several into the container to furnish aeration. Check the temperature and salinity of the water in the shipping box. If it is about the same as that of your aquarium, you can start transfer of the specimens in a short while, but if there is a great difference, then the procedure should be delayed until the temperature in the box is close to that of the aquarium. You can equalize both the temperature and salinity in the box by dipping up a glass of water from your aquarium and pouring it into the shipping box. Do this every few minutes until the temperature and salinity are well adjusted. This should also be done to adjust the specimens to the water in your aquarium, which in all likelihood is quite different from the water in the shipping box. Also, since your aquarium is full, almost to the top, you will need to remove several inches of water from it before you place the corals in it, or it will overflow. This is a standard procedure to follow with live fish as well as corals except that with live fish, you should not put any of the water in the shipping box back into the aquarium for fear of disease, unless you know the water is safe. With the live corals this is not as serious a problem so that you can return some of the water from the shipping box to the tank as you adjust the temperature and salinity. When you are satisfied that the temperature and salinity are equal, you may start transferring the specimens.

First, rinse your hands all the way up past your elbows, thoroughly, to remove all traces of soap or household oils. Then carefully pick up the corals and set them about the aquarium, placing the taller ones toward the back. Use a little imagination. Don't just plunk them down in rows. Nature never uses rows. Set the larger corals in back with some of the smaller corals nestled close to them, leaving some open spaces. The live rocks can be placed anywhere. Live gorgonians should be firmly anchored upright in the deeper sand. Live plants should have their root section pushed well down into the sand, and spread out so that their leafy portion does not touch the corals. Anemones may be placed in the open areas or on top of the live rocks. They will move about later to suit themselves. Crabs and snails can be placed in the open areas and these too will quickly move into habitat suitable to them.

Both the corals and live rocks should be turned so that the most interesting section is facing toward the front of the tank so that they can be viewed at their best. When all the specimens are in place, the aquarium should be filled to within one and one-half inches of the top or to the bottom edge of the top frame as explained earlier. The outside filter should be turned on and the aquarium light should be kept on, day and night for the next couple of days. This will aid the algaes on the rocks and corals as well as the plants.

If the corals give off a heavy scum when first placed in the tank, the

An aquarium with three small moon corals and several anemones as well as an assortment of live shells. This tank was kept outside in the bright sun. The specimens did better in sunlight than similar corals kept entirely inside. The corals should be gently placed in the aquarium so they won't be damaged.

No aquarium can be complete without a sea anemone. They are the "flowers" of the sea. I collected this beautiful specimen in just two feet of water. Anemones should be placed on rocks or open areas in the aquarium. They will move about to suit themselves.

scum may be removed with a fine filter net or siphoned from the tank with a length of plastic hose. As you are lifting the corals from the shipping box, it would be wise to examine each one briefly before placing it in the aquarium. Check it especially around the base for mold or fungus. If there is a white mold present, which will indicate decay, give the coral a brief "nose" test. There will be a bad odor if the coral is decaying and it should not be placed in the aquarium if a white mold or foul odor is present. Instead, return it to the shipping box and place it in a small, separate aquarium until you can determine its condition. Healthy coral will have a characteristic "ocean" odor that you will soon be able to identify after handling a few pieces. It has a rather strong smell but it does not have the powerful, nauseating odor of a piece that has gone bad.

Large chunks of coral or a large, living rock may be placed on end up against the back glass of the aquarium where they will make an unusual centerpiece, containing a whole community of life. I have

A large clump of moon coral that I kept alive in one of my aquariums. It lived for well over a year with little care.

collected many reef sections, fifteen to twenty inches across, and these are transported in one piece, filling up the entire aquarium. These would usually be dead rocks with a few chunks of live coral on them and would contain numerous plume worms, green algaes, small sponges, and a host of hydroids and tunicates. Other times a single piece of live moon coral a foot or two across would fill up the entire aquarium. People who saw it in my Miami shop were absolutely astounded. They couldn't believe it would live, yet many of the specimens were kept well past the year mark, which is quite a long time for a massive chunk of the reef to survive far inland.

If you collect all of your own live corals and fish and you are using natural sea water, you can add the fish to the aquarium two or three days after the coral has been set in place. I have done this many times with no problems. In fact I have often added both the fish and corals at the same time. It seems to work a little better if you put the corals and plants in first and give them a couple of days to adjust to the tank, especially the plants, which otherwise may be pulled up by the fish.

The spectacular Moon coral, which I named because of its resemblance to the surface of the moon, is one of the best corals for the aquarium.

The same coral with polyps extended tor feeding. A fan worm is visible at lower left.

A smaller head of moon coral. This specimen was only four inches across and easily handled.

The same coral with its polyps extended for feeding. This is one of the hardiest of all corals and one of the best for the invertebrate tank.

A fascinating assortment of marine life includes dozens of live seashells, hermit crabs, anemones, and the interesting alcyonium, a soft coral commonly known as dead man's fingers, similar to the gorgonians.

A small head of moon coral with a clump of dead man's fingers, Alcyonium, of interest to biologists, to the left of photo. Assorted algaes make up the balance of the aquarium.

But if you have purchased your corals, it would be best to let the live corals, rocks, and plants settle in the aquarium for at least three weeks to a month before adding any fish. This will allow the corals and invertebrates to become well established in the tank. The water will become aged, the undergravel filter will be operating at peak efficiency, and troublesome microorganisms in the water will by then have been filtered out of solution by the corals and mollusks (we hope).

The aquarium light should be left on for the first two days, then turned on for eight or ten hours a day or less if your tank gets a good amount of natural sunlight. In this case, it could be turned on in the evening for your own enjoyment.

If your corals are healthy and you have transported and transplanted them to the aquarium properly, you should now have before you a living coral reef, right in your living room. Turn on the light, get yourself a magnifying glass, and study this magnificent wonder in all its splendor. It should amaze you and you will see new things daily

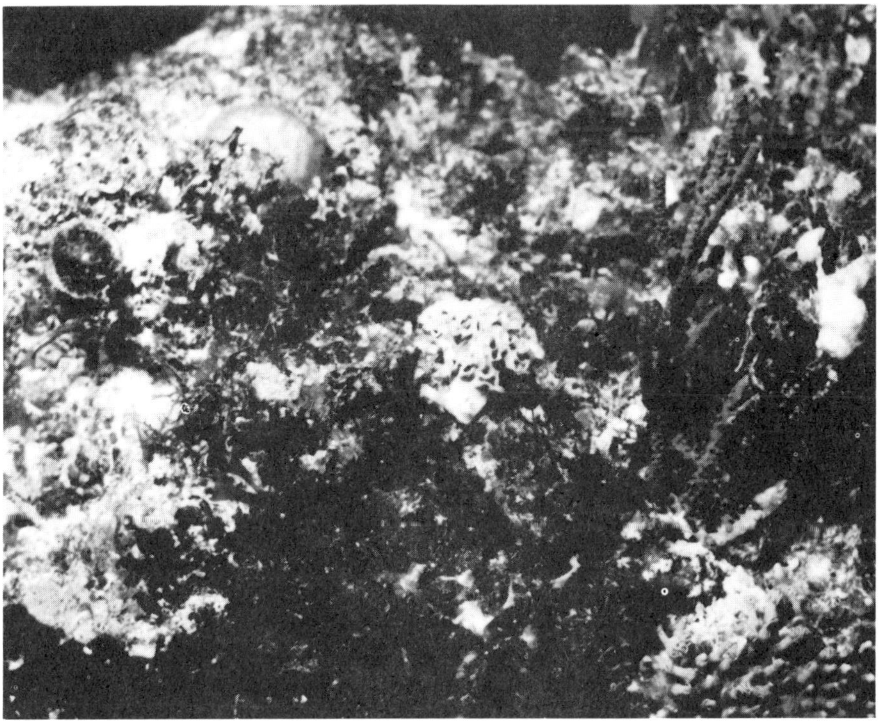

An "undersea" scene in one of my aquariums is simply a single large rock sixteen inches long and covered with interesting algaes and gorgonians. It can fill the aquarium and will bring forth a host of unusual creatures, such as the frilled nudibranch in the very center.

that you never knew were in your aquarium. A tiny brittle starfish may emerge from beneath a rock at feeding time. A live shell or dozens will crawl about the aquarium. Little mantis shrimp or even baby octopus may suddenly appear in the aquarium. There is really no limit to what can happen, especially if you get a wide variety of corals or several large "living" rocks. Some may even contain baby moray eels. This is what makes the live coral aquarium so fascinating. It is truly a part of the living sea, as complex as the ocean itself. Multitudes of marine creatures lay their eggs or hide within the crevices of the coral, and these will hatch out or appear perhaps weeks or months after the aquarium has been established. Your live coral aquarium will change and (hopefully) grow. You may see the corals reproduce; the plants will put out new shoots. Your fish may spawn, not once but many times, for they will feel as contented as if they were back in the sea.

A common sea urchin, center, and a club-spined urchin, to the right, make an interesting addition to the aquarium. Both are scavengers and will feed on dead tissue and decayed plant matter. This tank is using only a half panel undergravel filter. Note that it is pushed firmly into the corner. Bubbler stem gives off a strong surge of bubbles. This both aerates and filters the aquarium.

But you must use caution in introducing fish to the aquarium. There is a great amount of disease with coral fishes, most of it brought about by improper collecting and handling. If fish are collected properly, they should not have any diseases at all. I should know, for I spent a good portion of my life as a marine collector. The thousands of fish I caught and shipped all over the country rarely if ever came down with disease, for they were collected under the most careful conditions. I would never chill them or change them from one temperature of water to another, even on the boat, which is probably the biggest mistake made by many collectors. Another is that many collectors get their water and fish from different places. They may catch their fish on the clean, outer reefs, then gather their water near shore where it is highly polluted. Or they may mix fish and water from one part of the world with fish from another part of the world with the end result

You may keep colorful fish in your invertebrate tank such as this blue angelfish and neon goby with few problems if the fish are collected in a clean, unpolluted area. This tank contained many half-inch hermit crabs, which make excellent scavengers. Large hermits should not be kept in the live coral tank, because they will soon destroy the corals. Stringlike objects are Tetramin tube food.

that numerous microorganisms and bacteria are mixed together where they may multiply into plague proportions with no natural bacteria to keep them in check.

Consequently, unless you collect your own fish from a clean, unpolluted area, you should first quarantine the fish in a separate five- or ten-gallon tank for three weeks before placing them in your main tank. If the fish show signs of parasites, if they "scratch" or itch or dash about the aquarium, then the water in the quarantine tank should be medicated with copper or other medication until they are cured. Unless I collect my own fish, or personally know the collector who has brought them, I always medicate the fish with copper or copper and formalin while they are still in the shipping box. After they have been in the medication for a day, I then transfer them to the quarantine tank to observe them and give them additional medication if needed. Then I mix some of the water from the aquarium with the water in the quarantine tank, but I do not let any water from the quarantine tank get into the main tank. When I am ready to transfer the fish, I simply lift them out with a net and gently place them into the live coral tank.

The quarantine tank is kept in readiness so that if the fish are unhappy or contact disease in the coral tank, they can be immediately returned to the smaller tank for treatment. It is best to never put any medication in the live coral tank regardless of claims by the manufacturer that the medication will not harm the invertebrates. I have seen too many invertebrates killed by so-called harmless medicines. You will find that if your coral tank has enough live coral, anemones, and anemone rocks, many or all diseases that formerly plagued your fish will simply disappear. I have raised tiny angelfish and butterflyfish in live coral tanks without ever a sign of disease. They were the picture of health and they would find both a haven and a source of food in the live tank, for they would constantly nibble at the rocks and corals getting special food and minerals that they could not obtain otherwise.

This is why I feel that the natural, living reef tank is the tank of the future. It is far superior to the bleached coral tank in every way. In fact, if a collector or aquarium curator knows what he is doing, he will find that it is much less work to set up a complete live coral exhibit than it would be to get the corals, bleach and cure them, and start out the old way. He will find that the live coral tank requires far less maintenance, for the live corals don't need to be removed periodically for cleaning as with the bleached corals. In addition, even if corals die, they do not have to be removed except in extreme cases. If the corals are properly selected in the first place, and the aquarium is set up with the proper grade of sand and filtration, the death of a

Beachcombers along the beautiful shell-laden beach on Captiva Island. At low tide, many live shells and other interesting forms of marine life may be collected, right from shore.

number of coral polyps or even the entire coral head, will not usually foul the water at all. The numerous crabs and snails in the aquarium will feed on the dead tissue and the undergravel filter, that marvel of the aquarium age, will keep the water sparkling clean and fresh. I think if they had the undergravel filter a hundred years ago, the live coral aquarium would have been popular before the turn of the century, for the old aquarium literature records that as early as 1858, live corals and plants were kept in the marine aquarium with good results. In fact, one author even published information about which plants and corals did best in the aquarium and told how to collect them. I have not seen the literature but I suspect that it pertained largely to the northern corals and algaes. In addition, they did not have the all glass aquariums but used glass battery jars, which were good but which did not have the capacity in gallons of a regular aquarium.

I do not wish to give the impression that we know all the answers to keeping corals alive in the aquarium for we do not. We are really at the beginning of this fascinating field, on the very verge of discovery. Anyone who keeps live corals can learn new things about it and the day may come when all of the aquarium exhibits throughout the world will have living coral in their tanks, even in their huge circular oceanariums. From what I have already learned, I know their water would stay clearer, the fish healthier, the cleaning and maintenance less, and the overall picture of a coral reef would be presented in its true light, that of a living coral reef, brought inland for people to enjoy. Meanwhile, you can have your own coral reef right at home.

4
Feeding and Maintenance

THE EXACT DIET OF LIVE CORALS IS NOT KNOWN. WE DO KNOW THAT THE coral reef receives its nourishment from the ocean's currents, which constantly sweep countless tiny particles of plankton through the reef. As the tides rise and fall, different corals may feed and at slack tide, which occurs at both high and low tide, there is a period of a half hour to an hour when there is practically no current at all. At this time, especially during the slack at high tide, many of the corals will extend their polyps to feed. Anyone who has spent a great deal of time underwater on the reefs will quickly note this and, although in the past, many writers have stated that corals feed only at night, perhaps the bulk of the corals feed at the slack tide rather than the time of day. But corals do come out more after dark or even on a cloudy day. I have seen an entire reef feeding at high noon when there was a dense cloud cover and the tide was full. Even the basket starfish would spread forth its weblike tentacles to feed.

Plankton can consist of many different things, both animal and plant matter. There are zooplankton, phytoplankton, protozoans, copepods, radiolarians, medusae, plus larval forms of all kinds of fish and crustaceans. Plankton is really a soup of life and ultimately everything in the ocean depends upon it at some stage of its existence.

Live plankton would be the ideal food for the live coral aquarium but it is difficult to obtain at present. Frozen plankton can be obtained occasionally from tropical fish stores, and if this can be obtained, it should certainly be used. Meanwhile, we will need to devise a substitute that will keep the corals alive for a reasonable length of time. Artemia (live brine shrimp) is probably the best easily obtainable basic food. The brine shrimp eggs are available at any pet shop and they are easily hatched out in saltwater in just a day or two. Some

The secret to keeping exotic marine fish in the invertebrate tank is to obtain clean, disease-free fish. I collected this black angelfish myself and placed it in this tank of live rock and anemones. It never developed any signs of disease and a year later has more than tripled in size. Two six-inch lionfish were later added to the same aquarium with equal success.

brands are better than others, however, for some will hardly hatch out at all. The San Francisco brands seem to work best at the present time. They are hatched out by sprinkling a capful lightly on the surface of a couple of gallons of sea water. The water may be placed in a large fish bowl or a shallow, plastic container that is aerated slightly. It should be placed in the sunlight if possible and kept at seventy to eighty degrees. The eggs will hatch out into tiny wriggling shrimp in about two days. Usually the unhatched eggs will float on the surface or sink to the bottom and the tiny, pinhead-sized shrimp will be seen swimming about the container or a swarm of them will cluster around a bright portion of the hatching bowl. These may then be scooped up with a clean glass and poured into the aquarium. Or, you can use a large new oven baster and suck up a good portion of the shrimp as they congregate in a mass. This way you can miss getting the unhatched eggs, which will eventually become unsightly in the aquarium. If you

An all-glass tide-pool aquarium with algae-covered rocks and hundreds of invertebrates. The tank held several dozen grass shrimp, two dozen medium starfish, hermit crabs, live seashells, nudibranchs, oysters, clams, several dozen small anemones, sea cucumbers, and half a dozen blennies, perhaps five hundred specimens in all. Yet it contained only thirty gallons of water. Filtration was provided with sub-gravel filters and one outside filter with carbon. There was of course considerable predation, the starfish ate the oysters, the blennies ate the shrimp, but the tank never fouled and two years later it is still going strong. Note the Styrofoam and plywood under the glass bottom.

are a fastidious aquarium housekeeper and wish to keep your tank as spotless as possible, you can place the hatching tank up high, let it settle for five or ten minutes, then siphon only the live brine shrimp through a handkerchief placed over a large jar or bowl set at a lower level. This is the recommended procedure for sea horses or for most marine tanks and it is probably the best, but in the case of the live coral tank, perhaps the unhatched eggs will furnish gritlike food to at least some of the tank inhabitants. Either way is alright. If you wish to be extra careful you can siphon the shrimp through a handkerchief and then rinse the handkerchief in the aquarium, letting the shrimp wash down into the tank. Or if you aren't too fussy, simply dip up a glassful or two of the live shrimp, trying to avoid as many unhatched eggs as possible, and simply pour water and shrimp into your aquarium.

Usually there will be enough shrimp hatched from a capful for several days feeding. Make certain that you turn off your outside filter when feeding live brine shrimp. Otherwise they will quickly be sucked out of the aquarium. It may also be helpful to turn off your undergravel filters at feeding time, or turn them way down so they bubble very little. This can be accomplished by simply opening one of the unused air knobs on the gang valve and allowing the air to escape into the room. Or, if you like, you could insert an airstone into the aquarium, turning it on when feeding, while turning off all other filters. Keeping live corals is still strictly experimental and it is not certain which procedure will work best. You will need to experiment a little yourself. I sometimes pull out the air pump plug, as well as the light plug, and let the whole tank sit quietly for a half hour to an hour when feeding. This simulates the ocean at slack tide and seems to benefit the aquarium. But make certain that you turn everything back on after you finish feeding or your entire exhibit may be lost. If you

A stainless steel salad grater is best for preparing foods for the aquarium. Use shrimp, scallop, fish, clam, oyster, or similar food. The food should first be washed and frozen. Then it may be grated. The larger opening is good for anemones or large corals but the smaller openings are more desirable for the live corals.

A large gorgonian with its polyps extended for feeding in the aquarium. A feeding of newly hatched brine shrimp will usually "wake up" the gorgonian.

are forgetful, set the timer on the kitchen stove to ring after an hour to remind you to turn on the air pump or bubbler stems.

Feeding should be done twice a day, early in the morning, and just after dark, but if only one feeding can be done, it would be best to do it in the evening when you have more time to observe it. Live brine shrimp is by no means the only food you should give your aquarium inhabitants. They must have a great variety if they are to survive and grow. Feed them a pinch or two of a good grade of dry flake food such as Tetramin or other popular brands. Pulverize the food with your fingers and swirl it into the water. If it floats, it will eventually sink when it gets over the bubble stems, for the surge of the water will wet it and help to distribute it about the aquarium. If you like, you can mix an eighth teaspoonful or less of the dry food in with the live brine shrimp, wetting it thoroughly so it will sink, and feed it along with the live shrimp.

The best results in feeding will come with a great deal of experimenting. An ordinary stainless steel salad grater or shredder, the type with a variety of sizes and openings, will be excellent for preparing food for the aquarium. You can start out with a chunk of frozen fish, one or two frozen shrimp tails, and a frozen clam or oyster. Using the finest opening on the shredder, carefully shred the fish, shrimp, and clam (or all three of them together) into a shallow bowl. Then add water and stir gently. Pour contents into a quart glass measuring container and observe it to see that it is finely shredded. Then very slowly pour off the water until you come to the shredded matter, which should settle to the bottom. If the water is very milky, make one or two rinses until the water is clean. Don't try to get it too clean, for you want some fine suspended matter in it, but you do want to get out the bulk of the white juice because this can quickly pollute the aquarium. Then, stir in a small amount of fine, baby dry fish food. Or, if you can obtain it, shred in a small amount of lean beef heart. You will now have a basic coral food and this can be fed sparingly to your corals. Feed it at night or in the early morning, turning off your outside filters or all filters as you would with the live brine shrimp. This will signal your coral that it is time for dinner. Turn off the light also. Then, when the coral polyps start to emerge, swirl a quarter teaspoon of the macerated food down into the tank so that it lands on top of the corals you wish to feed. You may find an oven baster helpful because it will enable you to direct the food to the specific corals you wish to feed. You may also sprinkle baby dry fish food on the surface and let it settle naturally on the corals.

You will notice that after a few feedings, many of the corals and gorgonians will open up at the first sign of food in the water and will

A live pen shell on the right is open for feeding. It filters fine food particles from the water. I collected it and placed it in the aquarium, where it lived nearly a year.

begin feeding. Some of the corals, such as the moon corals, have long tentacles on their polyps and will hang these out a half inch or more to collect food. The gorgonians will grasp the food with their eight tentacled polyps and the corals will emerge like little pussy willows. From beneath the corals, long-armed brittle starfish will quickly emerge, tossing one arm above the next in their anxiety to reach the food before it is all gone. Some of the brittle stars will become so tame that they will actually take food from your fingers. Octopus will do this too as will most of the crabs, especially the long-armed arrow crab. He will spot the food clear across the aquarium and start reaching for it.

Feed the shredded fish and shrimp sparingly, for too much of it can quickly foul the aquarium. Spread small amounts of it directly on the corals and observe them to see if they eat it. If you see that they don't like it, you may need to change the ingredients. It sometimes works best to shred your food into seperate jars, labeling each, and feeding them to the different corals seeing which does best with each type of

Giant clown anemones such as this stoichactis are a favorite for the invertebrate tank. They are very active, and when inhabited by their symbiotic companion, the clownfish, they present one of the marvels of the world. But keeping them is not easy. Apparently they need prodigious amounts of food to sustain them. Copious feedings of live brine shrimp, fish, and shrimp are needed. Also good light and perhaps some sunlight. Treat them much like corals.

Flame scallops *(Lima scabra)* often live within the coral and the feeding of them is the same as the coral. This specimen is open wide for feeding.

Closeup of alcyonium with polyps extended. These are extremely hardy in the aquarium. A simple magnifying glass will allow close study of the living specimens.

food. In addition to the shredded food, there are dozens of prepared tropical fish foods that you can try. Some contain algae, plankton, vitamins, and other healthful ingredients. Some aquarists pulverize a rabbit food pellet or alfalfa pellet into the aquarium from time to time as a food supplement, with good results.

Fresh fish may be an especially good food for it is one of the basic foods of the sea. The "big one eat the little ones" all the way from the whales on down to the sardine, so that fresh fish is likely to prove one of the best foods. A small fillet of fish will provide many scrapings or shreddings for the coral tank. Keep it well wrapped in the freezer to avoid dehydration and feed it sparingly as explained above. But vary the diet with other foods as well. Vitamins should also be tried, especially vitamin C, and multi vitamins. Use liquid forms and mix with the food.

Coral polyps are very much like sea anemones except that they live in a hard limestone or calcium carbonate tube. They have nematocysts in their tentacles and can sting their prey. They have even been

A live coquina *(Donax variabilis)* clam with siphon extended, indicating that it is feeding. These colorful little shells are found right at the surf line and are easily collected by picking them the instant the surf retreats. Note the dozens of siphons protruding above the sand to the right.

known to kill fish and can devour a small fish completely if it becomes entrapped in their tentacles. In one of my books, *Exploring the Reef,* there is a photo of a live coral actually devouring a fish and to my knowledge, this is the first time that this has ever been portrayed in any book. Of course coral eats tiny fish routinely as plankton but it can also devour larger chunks of flesh, perhaps eating remnants of fish that were chewed and devoured by sharks or other predators and that has settled onto the coral. In the ocean, very little is wasted. Coral has a rather catholic taste and due to the complexity of the sea, the precise diet can hardly even be guessed at for the ocean currents contain a multitudinous and heterogeneous variety of particles of every

Live coral devouring a fish in the aquarium. I was perhaps the first to discover that coral will devour fish and reported it in my book *Exploring the Reef,* in 1968. It was later substantiated by Robert Sisson in *National Geographic.*

known mineral and element as well as the planktons. It is virtually impossible to even analyze it. At this stage, we can only furnish a rudimentary diet and hope for the best.

But as crude as it is, the aforementioned foods will sustain at least some corals for a year or longer. Feed your corals sparingly with the fresh fish or clam and more heavily with the live brine shrimp and dry foods. Make certain that the aquarium contains a good many scavengers such as sea urchins, starfish, dozens of tiny hermit crabs, and live sea snails. These will all help to clean up the bottom. If you are fortunate, not only will your coral live, but it will actually grow! Most corals grow rather slowly, adding about an inch a year, depending upon the species, but little or nothing is known about their growth in the aquarium. Think how exciting it would be to watch a piece of elkhorn coral (*Acropora palmata*) grow a new limb. This would be a first in the aquarium field and yet it is entirely possible. Practically everything you do with a live coral tank will be a first. No two tanks can ever be exactly the same. Your tank might hatch out a rare species of seashell. Wouldn't it be fantastic if you had a piece of coral that contained the eggs of the rare slit shell? If they hatched out and you could raise them it would be the greatest achievement in malacological history and you would be famous and perhaps even rich considering the fact that the shells are worth about five hundred dollars each at this writing. Now wouldn't that be something? Yet it is entirely possible. When it comes to the sea, anything can happen, especially when the deep reefs are explored. Here we have a wide open field of discovery for the aquarist. Who will be first to keep live coral from the great depths?

I had some of the deep-sea coral brought to me by Dr. Paul A. Zahl while assisting him with his article on "Fluorescent Sea Life" for *National Geographic* (August 1963). He has collected it off the East coast of Florida from a deep-sea scallop dredge and brought it back to my laboratory in Key Largo Florida for study. After he finished his project, I kept the coral alive for about six to eight months, with no trouble at all. The coral was snow white with pink polyps. Apparently the deep-sea corals require much less food to survive than the shallow corals, for there is very little suspended matter at the great depths, the water being nearly totally clear. Consequently the animals at that depth get little food in comparison to those near the surface and have learned to survive with less food.

Aquarists throughout the world will undoubtedly devise and improve their own special diets for corals so that the keeping of them will be greatly extended. But in addition to food, most corals will need a good deal of light to maintain the algae and the single-celled plants zooxanthellae that live within the coral polyps. These play an important part in the survival of the corals themselves. There is dispute as to how the coral actually utilizes its foods, whether or not it feeds on the zooxanthellae or whether the zooxanthellae subsists on byproducts of the coral polyps, but this will be for scientists to decide. The corals do need light of some type which is why most of them are found near the surface or at most a few hundred feet down.

It stands to reason that if you find a piece of coral in shallow water a few feet deep and it is out in the bright sun all day long, then that coral will need a good deal of light in the aquarium if it is to be kept in good condition. On the other hand, if the coral comes from the fifty- or hundred-foot depth, it will require much less light. Unfortunately, the deeper corals are most difficult to obtain at present so we must assume that our corals are from shallow water, and we should give them a good deal of light or some actual sunlight if at all possible. Someday I hope we can get some of the solitary corals from the thousand-footh depth or even the ten-thousand-foot depth, for I feel that

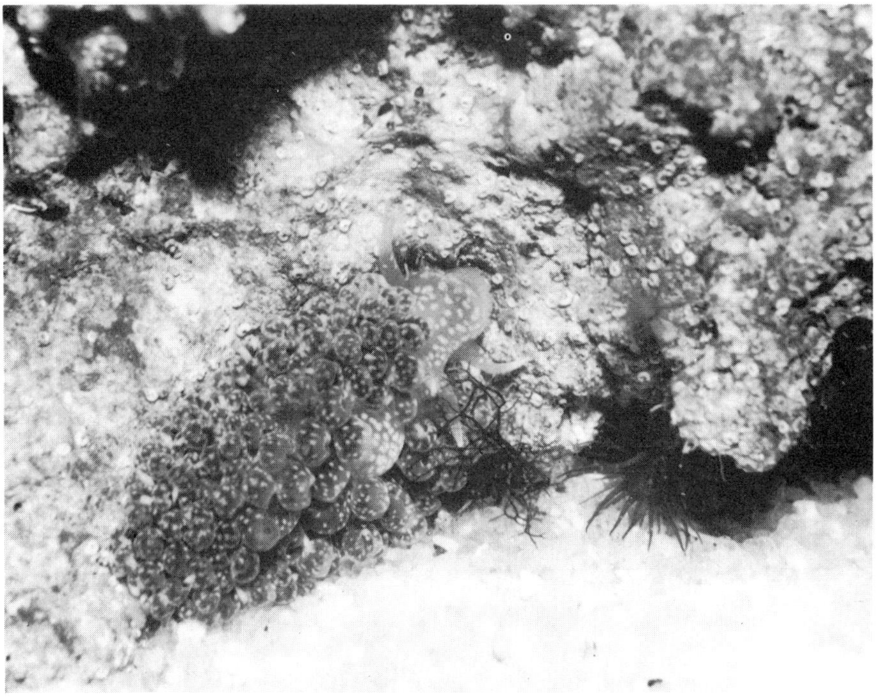

A simple algae-covered rock will provide food for a variety of creatures. This frilled nudibranch is feeding on various growths.

The same nudibranch feeding on dwarf anemones. It is difficult to determine exactly what each marine organism will eat, but by placing a large algae covered rock in the tank, it will often contain enough invertebrate life to furnish food for a variety of specimens.

Excess food should be swept from the tank bottom with a fine silk filter-net. Shredded shrimp or fish in particular can easily foul an aquarium if fed in excess. The filter net will remove the food by simply swirling it just above the sand.

these would live well in the tank, although refrigeration may be needed. It should be pointed out that although coral reefs in general are found no deeper than a few hundred feet, some reefs have been found at a depth of a thousand feet according to *Sea Frontiers*, from the University of Miami. This would be a whole new field for the live-coral hobbyist or student if he could obtain specimens from that depth. In addition, some isolated, solitary corals are said to be found at even the greatest depths of the sea so the researcher is not likely to run out of specimens to study.

Regardless of which type of corals or live rocks you are keeping, you will need to experiment with feeding them. Feed sparingly but give them enough. Sweep up excess food with a fine nylon "filter net" if there appears to be an excess on the aquarium bottom. Give your fish, crabs, and shrimp larger chunks than the fine shredded food, since this may be too fine for them, but don't overfeed or you will have bacterial or parasite problems.

Things To Watch For

Check all of your corals very carefully the first few days for signs of decay. If you notice a white fungus or mold near the base or side of a coral, remove the coral promptly. This may be moldy decay and is a sign that the coral is dying or dead. The coral will have a very bad odor as explained earlier and it should not be left in the tank because it may kill the other corals. Sometimes you can wash off the decayed portion of the coral by holding it briefly under a strong jet of water, washing only the affected part. If the rest of the coral appears to be in good shape, it may be returned to the tank but it should be closely watched to see that the condition does not spread. If it is in bad condition or questionable shape, it would be best to keep it in a separate tank after first washing off the decayed area. Should it eventually recover, it could then be returned to the main aquarium.

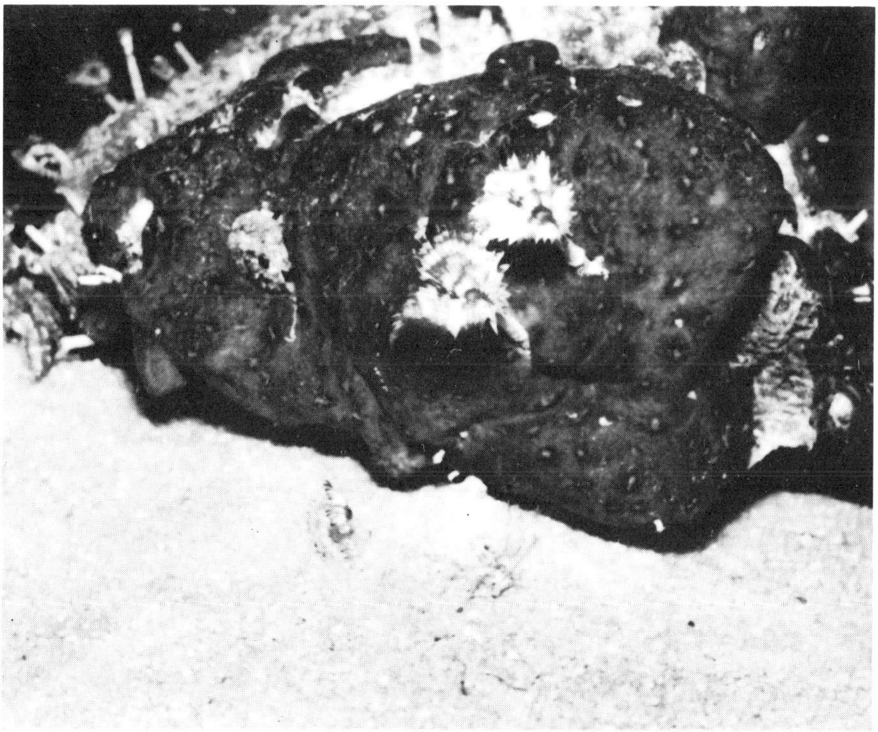

Be especially careful if coral or live rock contains black or brown sponge like that shown above. This is difficult to keep alive and if it dies it will usually foul the entire aquarium. The author picked this rock because it housed dozens of interesting tube worms. Note that the two in the center have "shed" their plumes, a fact I discovered with my aquarium. This would have been nearly impossible to discover under the sea.

Be especially careful if your coral contains brown or black sponge. This will appear as a smooth, unbroken area, usually near the base of the coral. It is best to obtain corals with no sponge at all, for the sponge usually dies in the aquarium and can easily foul a tank. If a portion of the coral appears to be greatly swollen with gas bubbles or gives off "smoke" when prodded with an aquarium scraper, then it may be decayed. Remove the coral and smell it. If it smells like ten thousand dead skunks or worse, then it has gone bad. It will usually leave a trail of cloudy water as you lift it from the tank. It should be promptly removed from the tank. If it contains live plume worms or mollusks, rinse off the decayed area and then place it in a separate bowl or aquarium for observation, returning it to the main tank a week or so later if it cleans up.

Observe the clarity of the water. It should be crystal clear. If it is milky or cloudy, then some of your corals are dead or dying. Check each one visually at first and then use the famous Straughan "NOSE TEST," which is the most accurate method of determining the condition of a piece of coral. Lift up each coral and sniff it briefly. If the coral is dead or partly dead, your nose will quickly discover it. Check your filters to see that they are operating properly and pick up any uneaten food from the bottom of the aquarium. If the water is very murky, it would be wise to change a portion of it, perhaps as much as a half. It is a wise procedure to always keep ten or twenty gallons of clean sea water on hand for such an emergency. You can store it in a closet or in five-gallon glass jugs, under the aquarium. Gallon glass jugs are excellent for storing water, and the gallon, plastic milk containers will work fine.

Live gorgonians, alcynarians, or sea whips should be observed for signs of life. Usually they will put forth their polyps to feed after a day or two. But if they show no sign of life after a week, they should be checked to see if they are alive. The nose test works with them also. In addition, if they are dead or dying, they will give off a milky cloud when prodded. Also, some of the body will come loose, revealing the black or dark brown skeleton inside. If this is the case, they should be promptly removed. However, if they appear firm and have no noticeable odor, other than their fresh, natural odor, leave them alone and perhaps they will emerge healthfully in a week or two. Some corals, such as the orange polyps coral or deep water corals, may not emerge at all for several weeks. They will appear to be dead until one day you will find the whole aquarium alive with brilliant blossoms. This is just one of the surprises of a live coral tank. I had this very thing happen with a large shipment of live coral from Puerto Rico. The coral had arrived in huge clusters in very dirty water and it appeared

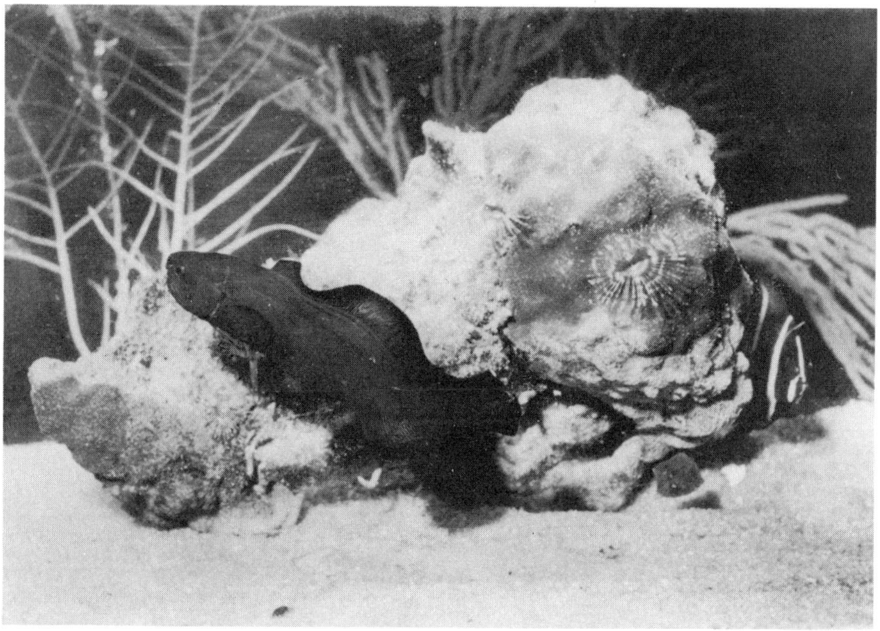

With the exception of sponge, which will foul the water, most corals may die a little at a time and never foul the tank. Some specimens, like the one above, may be partly dead when collected, which is evidenced by the white area at the top. This was buried in the sea floor yet the coral flourished on the exposed areas. The unusual fish is a brotula, sometimes called the Black Widow.

to be dead. But it had no bad odor so I set it up in a number of aquariums after carefully rinsing each piece in clean sea water. Weeks went by and it showed no sign of life, until one morning when I visited the shop, I found that "spring" had come to the aquariums, for the coral all came out in full bloom with brilliant orange blossoms. It was a spectacular sight. After that, it blossomed nearly every day, especially at feeding time. So don't throw out the coral unless it smells decayed or is positively dead. Even if it is dead, if it has no foul odor, new life may spring from it after a period of time.

Some corals may die a little at a time. Either the fish will kill the polyps, or crabs may devour the tissue. This is no cause for alarm except that you should watch the dying area closely. If you can see which fish or crab is doing the damage, remove the offender to another aquarium or isolate it in the corner with a sheet of glass. Usually, if only a part of the coral has died, this doesn't mean that the whole head will die unless conditions were not right in the aquarium. It should be remembered that butterflyfish, parrotfish, and certain filefish feed on coral so these should not be kept in the tank unless you are trying to rear them and can spare some coral as food. If small portions of the coral turn white and appear to be dead and the white color is caused from removal of the tissue rather than a white mold, then it will be safe to leave the coral right in the tank, since this condition is not as serious as the white mold. Sometimes a portion of the coral is damaged or fish, crabs, and invertebrates may eat some of the polyps exposing the white coral. This may heal up in time and even be overgrown with fresh coral if you feed and tend your aquarium properly.

Live Rocks

Since these usually contain no coral polyps, they don't usually die as quickly or as noticeably as coral. But if they contain brown or black sponge, this will rot and foul the tank unless promptly removed. If there is sponge on either the coral or live rock, and it dies, remove every trace of it from the rock or coral. It can be peeled off with a knife or with the fingers. Once the coral is rid of it, it can cause no further problem. Things to watch for with live rocks in addition to sponge growth are the white mold and decay from within. If a large clam or mollusk inside the rock should die, it can cause problems. Again the nose test will prove useful in checking the live rocks. If your aquarium water is crystal clear and there are no signs of decay or smoky clouds rising from the rocks, most likely they are alright. Another way to tell if all is well with your live tank is to smell the surface of the water each day as you check the aquarium. There should be no noticeable odor except for a slight ocean smell. If the tank has a strong earthy smell or a foul odor, then something is rotten, and it won't be in Denmark! A nitrite test kit may also prove useful.

It should be pointed out that not all sponge is bad. Some types will grow on the live rocks and thrive in the aquarium. But the smooth types that resemble smooth leather usually will not make it. Neither will the bright red types or other bright-colored growths from the deeper water unless you change a good portion of the water each day. If this is done you can keep practically anything alive indefinitely.

Live Plants

Most live plants or algaes will do well when conditions are right. If the plants are fine and healthy, they will be bright green, at least in the daytime. If they turn white or brown and the light is good, then they have probably died. They should be removed from the tank. Some of the plants like the Penicillus or udotea will put out new

Live plants (algaes) will add beauty and interest to the invertebrate tank and do quite well in the aquarium. I kept and shipped them over twenty years ago while most books said they wouldn't live in the aquarium at all.

plants if the dead stem is planted just beneath the sand. Remove all of the top, fleshy, loose material and plant the remaining stem in the tank. If you are fortunate, you may see new plants growing there after a few weeks. Plants that have long runners or rhyzomes, such, as the caulerpas, should be obtained in one piece and should not be broken apart. Many of them will "bleed" and lose their body fluids. Also the root sections should be treated carefully and well anchored in the sand. Although some scientists feel that the roots of all marine algaes are merely "holdfasts" and contribute nothing to the plant except to serve as an anchor, I disagree. Some algaes have well-developed roots and if these are not planted, the plant will quickly die, so at least some algaes appear to have genuine root systems. The caulerpas fall in this category and have what appear to be well-developed root stems as complex as any land plant. Plants that have been uprooted by fish or crabs should be promptly replanted. Some plants, particularly the various species of caulerpa, will grow and spread throughout the tank. If this happens, leave them alone, except to replant unrooted sections. Large leaves that have died and turned brown or white may be removed from the tank unless there are sufficient snails and urchins to devour them. There is still a great deal to be learned about plants and you

In addition to their beauty, marine plants condition the water, making it a healthier environment for the specimens. It was with the use of live plants above that the first mass spawning of pygmy sea horses took place in the aquarium. More than a hundred sea horses spawned. I reported the event in the *Aquarium Journal*.

as an aquarist can contribute your share by keeping records and trying new algaes and methods. Give the plants good light, preferably sunlight, and they should do well.

General Aquarium Cleaning

Once a week, or whenever it needs it, the inside front glass of the tank should be cleaned. A sixteen-inch aquarium scraper will prove useful here. The type with a cleaning sponge on one end and a plastic scraper on the other is the most useful, for the sponge can be used to wipe down the inside of the glass. The scraper will also be useful for removing stubborn algae from the glass as well as for moving sand or small items about the aquarium, reducing the need to place your hands in the tank. The less you put your hands in the tank the better, so the above gadget will prove useful in many ways.

The penicillus or merman's shaving brush is a popular plant. It must be obtained in good condition if it is to grow. The root or "holdfast" should be imbedded in the sand. Good light is also required.

One of the most beautiful of marine algaes is acetabularia, sometimes called mermaid's bra or mermaid's umbrella. It is a delicate plant and needs strong light or sunlight. A scientist in Miami from the university raised acetabularia through three generations on the roof of the laboratory in pure synthetic Reef Salts solution. Dark objects are sea slugs.

If algae is growing on the glass, you may wish to leave it grow on the back glass or on the sides, in which case only the front glass should be cleaned. The bottom of the tank should be checked for excess waste matter and if there is an accumulation of it around the corals, it should be siphoned from the tank. It doesn't hurt to leave a little of it in the bottom of the tank, for the scavengers to eat. Little hermit crabs, sea cucumbers, and snails will devour much of the debris. Any large chunks of food that remain uneaten should be removed. Do not think that your "biological filter" will solve all of your problems and cannot be overtaxed. Some aquarists think that once a biological filter is established it can handle any condition that may arise in the aquarium. But this is not so, for everything has its limits and the undergravel filter is no exception. The death of a large conch or cowrie or a single large gorgonia could easily overtax the system and turn your once-beautiful living reef into a soup bowl of sticky okralike chowder. It will smell so bad that it may go as high as 8.9 on the Nose Scale, which is the equivalent of two hundred skunks and a dead horse at high noon! Nothing smells quite as bad as rotting coral or gorgonians. As you attempt to clean out the nauseating mess, your nostrils will quiver in pain, your eyes will water, and the entire household will complain. Even the neighbors will get the odor.

Twenty years ago, I set up a saltwater tank for some people in Miami and put five large four-inch cowries in it, figuring it would make an interesting display. The cowries lived fine for a while and I didn't go back for several weeks. When I did, I noted that they had died and were lying in the bottom of the tank. The water was a soupy brown but had developed a heavy film on top, keeping the odor confined. I decided to empty it. I knew there would be some odor so I decided to get the messy task over with quickly. I picked up the tank, water and all (it was a large glass battery jar), and dumped it in back of the house. WOW! Such an odor I have never seen. It shot through the house like a cloud of mustard gas. The people were eating dinner and rushed back to see what had happened. There were loud exclamations in Spanish, which I could not understand, but I gathered that they were not happy with the smell, especially at mealtime. I got some of the slime on my hands from the shells while attempting to clean out the shells and the horrible odor persisted on my hands for days despite intense soaking and scrubbing with soap and water.

Don't take a chance on offending your neighbor's nose with a tank of live coral that has gone bad. Instead, select each piece of coral that goes into it with care and check the tank each day for signs of decay. If you take out a dead piece promptly, the trouble will stop there, but if you allow a large piece to rot in the tank overnight, it could kill all of the other corals, which would be sad considering all the work you did to get the tank set up. Purple gorgonians or sea fans in particular can cause a serious problem. Make certain they are in perfect shape before you add them to your tank. Sponges also can be deadly. Don't put any in the tank at all unless you are positive they will live in captivity.

Sea anemones generally will live well and cause few problems unless you overload your tank with them. The large Pacific clown anemones in particular should be watched very closely, especially for the first few days. If they don't fill out within a day or two, they should be removed to a smaller tank and kept by themselves until they appear to be alright. If they stay on the bottom in a clump and give off a cloud of "smoke", then they are probably dead and should be removed.

Changing the Water

Your live coral tank will generally do better if you change some of the water each month. If you live by the sea, you can change a third of the water each week by simply bringing two buckets of water into the house and adding it to the tank after first removing two buckets of water from the aquarium. But if you are an inland aquarist and water is at a premium, you will need to be more cautious with it. In this case, you can remove a gallon or two of water from the tank each week and replace it with fresh sea water or artificial sea water that has been mixed the month before and allowed to age. Use a gallon measuring cup and take the first scoop from the top, tipping the scoop slightly into the tank so that you scoop up a good portion of the surface film. This will help clean the surface of scum that may form there. Then, lift up some water in the scoop and pour it down into the tank, directly on the coral or rocks. This will help to flush them of loose skin in the case of the corals and accumulated dirt in the case of the rocks. Don't pour the water from too great a height because it may damage the coral. A foot above the surface or less is sufficient. I have discovered that at least some corals appear to shed their skin. This has been observed on the reef and in the aquarium many times. The coral becomes covered with slimy film. In stormy weather, the entire film is sloughed off by wave action and strong seas. The same thing can happen in the aquarium, except that *you* must make the wave action once or twice a month by pouring water over the coral. After you have gently flushed the coral, remove a gallon or two of water and replace it with new sea water. Do this once a week or at least once a month. You should replace about four inches of water from your tank each month if possible. But this is a flexible figure. It can be more or it can be less depending on how many specimens you have in your tank and how well you tend your tank.

Outside filters should be emptied into a bucket and cleaned at least once a month. The carbon should be dried out in the sun if possible, or baked in a Pyrex baking dish in the oven until dry. You can revitalize the carbon a little if you will crush it in your hands while washing it. Squeeze it tightly, rubbing it together to pulverize it slightly. This will clean its outer absorbing surface. It should then be rinsed in clean water until it is sparkling clean.

The sub-sand filter can be left entirely alone with the live coral tank. Except in rare instances where the tank is thoroughly polluted, you can leave the undergravel filter alone for years, disturbing it as little as possible, except to clean around the corals. The sand will become the home of numerous sea worms and minute forms of life, much the same as the sand in the ocean, and this minute life will greatly aid in keeping the tank in prime condition. I have taken down coral tanks to move them in which the same sand had been used for nearly five years and yet the sand had no noticeable odor at all, thanks to the bacterial action, filtration, and circulation of the water through it from the undergravel filter. Multitudes of tiny filter feeding bacteria, tube worms, and sand dwellers all worked together to keep the aquarium healthy.

The sand around the corals may be cleaned once in a while with a fine silk type or fine nylon filter net. This is a very fine netting that will trap even the smallest dirt. It is swirled about the tank just above the sand and it will actually draw dirt right out of the sand. I have recommended its use in the marine aquarium for many years and it can do a fantastic cleaning job, eliminating the need to ever break down the aquarium except in rare instances. Sometimes it will benefit the aquarium and the corals if you will stir the sand very lightly. This will allow a certain amount of sand and dissolved material to be ingested by the corals, but don't overdo it or you will destroy the filtering action of your undergravel filter. Try it just a little from time to time and evaluate the results.

Don't forget that you must NEVER PUT ANY MEDICATION IN YOUR INVERTEBRATE TANK. If your fish or specimens need medication, simply remove them and treat them in a seperate tank. There are some medications containing copper that are claimed to be safe for invertebrates, but personally I would not put them in any of my invertebrate tanks. Sometimes just because a medication doesn't kill a hermit crab or coral shrimp right away, the aquarist thinks that the medication is harmless, but a few weeks later he may find that most of his invertebrates are dead or dying.

Sunning your coral. If your aquarium has no sunlight, it will benefit the coral if you will place it in a bowl outside in the bright sunlight at regular intervals. Should your neighbor inquire about your activities simply tell him you are "sunning your corals." He will understand.

Sunning Your Corals

If your aquarium does not get any sunlight, it may prove worthwhile to give your corals a sunbath, at least once a week. Simply dip up some water in a large bowl, place one or two corals in it, and then set it out in the bright sunlight for an hour or two. If anyone should ask you what you are doing, just tell them you are "sunning your corals!" I am sure they will understand. Seriously though, this will definitely help your corals or algae rocks. Stay out there with them and don't leave them in the bright sun for more than a couple hours or the water may become too warm, especially if you live in the southern states and it is summer. In cooler weather you could give them a longer sunbath if time permits. Not much is known about sunbathing coral at this time. I have tried it a number of times and it seems to help, but it will take a great deal of experimenting to determine if it is really worthwhile, as it appears to be. The water in the bowl should be poured back into the aquarium when you return the corals to the tank. It is well known that corals need sunlight, and I kept some of my corals outside in the bright sun by simply setting the aquarium up against the house where it will get the sun. The corals did much better outside than they did in the house in nearly every instance so that sunbathing may be practical if you can't keep at least a small tank in a window or outside. Anemones, too, may require an occasional sunbath.

Checking the Salinity

The salinity or specific gravity of the saltwater aquarium should be checked once a week or at least twice a month. Saltwater evaporates surprisingly fast in the aquarium, especially if the home is air-conditioned or if the weather is very hot. A hydrometer made for measuring saltwater should be used, and these are available at most tropical-fish stores. They usually have two red lines about a half inch apart and the accepted reading of 1.025 is in the center, between the lines. Larger

Sea anemones will take much of their food as it swirls through the water but generous portions may be placed directly on their tentacles. They may also require some sunlight.

hydrometers a foot in length, although much more expensive, are easier to read because of the larger scale and are usually more accurate. A scale of 1.000 to 1.050 is best and this will cover the extremes in salinity from fresh water to very salty. It is best to keep the salinity slightly on the fresh side rather than on the salt side, so a reading of 1.024 is fine.

It should be remembered that the salt usually stays in the aquarium as the water evaporates unless you have heavy aeration with air stones, in which case some of the salt will dry and collect on the cover glass. This can be scraped back into the tank if it is clean and not contaminated by dirt or rust from the aquarium light. Some of the newer aquarium lights are all plastic instead of the stainless steel of the past, and these should be used for saltwater because they are far superior. To take the reading with the hydrometer, simply insert the instrument in the aquarium, dislodging any bubbles from it, and when it has stopped all movement, look directly at the water line, which is where the reading is taken. If the reading is higher than 1.025, say around 1.027 or 1.030, then it means you must add fresh water, for the tank is too salty. It sometimes takes a gallon or two of fresh water to bring it back to normal depending on the capacity of the aquarium. Rainwater, distilled water, deionized water, or well water is best, but if none of these can be used, tap water will be alright. It should be placed in a gallon jug for a week before use to allow chlorine to escape and to allow minerals to settle. It should be poured very slowly into the aquarium to avoid stirring up sediment from the bottom of the jug.

By filling the aquarium exactly to the bottom edge of the frame at the top of the aquarium, you will be able to see evaporation immediately, for as the water evaporates it will drop below the aquarium frame. You will then know that it is time to add fresh water to the tank, but check it with the hydrometer to be certain. If you are using a small hydrometer, you can keep it in the tank, especially if you are using a hydrometer-thermometer combination. Wipe the hydrometer with a paper towel before taking the reading to remove surface scum or oils. If your aquarium doesn't have a frame at the top, make a mark one and a half inches from the top at the side of the tank and keep the tank filled to that level.

Cleaning the Outside of the Tank

Saltwater is very corrosive. If any spills outside the aquarium and gets on your tank stand, it will quickly rust it. Use rubber or plastic coasters under the stand legs to keep them from rusting the floor. Wipe up any saltwater immediately should it spill outside the aquarium.

The outside aquarium glass should be rinsed lightly with a clean sponge using fresh water once or twice a month to remove any traces of saltwater. Then the glass should be polished dry with paper towels or lint-free cloth. This can make a big difference in the beauty of your aquarium. If saltwater is smeared on the front glass and wiped off without first rinsing with fresh water, it will give the glass a dull finish and the attractiveness of the aquarium will be diminished. Many public aquariums do this, hiring someone to wipe the tanks of fingerprints but failing to instruct them that they must first rinse the salt from the tanks. As a result, although the tank itself may be sparkling clear, it will appear dingy simply because the glass is not really clean.

Adding New Corals or Specimens

New corals, plants, or fish may be added at any time once you have mastered the general procedure of keeping them. The new corals or fish should be placed near the aquarium in their shipping box and water from the tank mixed with water in their box until temperatures are equal. If the water they arrive in is artificial and you fear it may harbor disease or parasites, don't put any of it in your aquarium, but if it is pure, fresh seawater, you can add the water to your tank after you have adjusted the temperature and salinity. Older corals that have died and are no longer of interest may be removed if they show no signs of algae or tubeworms sprouting forth, to make room for the new specimens. On the other hand if you have been extremely successful and your corals have grown a great deal, you might have to remove a few pieces to another tank to give them room to grow. If this happens consider yourself an unqualified success in the keeping of live corals in the aquarium.

5
Collecting Your Own Specimens

IF YOU LIVE BY THE SEA, THE LIVE-CORAL AQUARIUM WILL BE A SNAP, FOR you can experiment with all types of interesting corals and plants with little worry or expense. In fact, if you live by the sea, the live-coral tank is the only way to go, especially if you live in the Bahamas or Caribbean.

I once visited a scientist in the Virgin Islands who lived right by the sea next to a magnificent coral reef. Imagine my surprise when I found that he was keeping fresh-water fish in the aquarium in his living room. Right at his doorstep could be found royal grammas, rock beauties, and all kinds of exotic fish, not to mention the many live rocks and corals. Had he only known that a saltwater tank could be set up so easily, I am certain he would have had one, but I suspect the thought never occurred to him and he went on with his guppies and swordtails, never knowing what he was missing. What a shame! What a pity! He could have enjoyed the wonders of the living sea at no expense other than picking up the specimens at his doorstep.

Nothing is as fascinating as a saltwater aquarium, except two of them or a good looking blonde. I have set up saltwater tanks for millionaires and the heads of giant corporations and yet with all their wealth, they would be absolutely entranced with the wonders of the sea. A simple barnacle rock with pulsating "hands" swirling through the water would be utterly fascinating to them. A plume worm disappearing into the coral would bring a scream of delight. A tiny hermit crab emerging from beneath a coral to grab a tidbit of food at mealtime or the wondrous red and white coral shrimp that looks like a living candy stick is all part of the undersea scene and far more interesting than any television show.

Those of you who are fortunate enough to live by the sea should bring a little of the sea indoors. Set up a fifteen- or twenty-gallon tank or several or a six- to eight-foot tank if space and finances permit and you will have a continuous undersea show that you can enjoy every day regardless of the weather. It will bring you more delight than you ever thought possible and it is very educational besides. You can learn more things about the corals and the reef from the comfort of your livingroom than you could ever learn under the sea, for you can observe it around the clock and examine it with a magnifying glass if you like. This would be difficult to do under the sea unless you were living in an undersea chamber. Even then, you couldn't get as close to the corals as you could viewing them through the glass sides of an aquarium.

All you have to do is get a few buckets of water to fill your aquarium, gather up some nice clean sand if you live in an unpolluted area, and you are in business. Get your sand on the coarse side, from a rocky area where the sand has washed high up on shore. Use an aquarium fish net to gather it and rinse out the fine sand. Then rinse it out in a bucket until it is clean as described in chapter 2 and you are in business. Fill the tank, get the pump and filters going, and then go out and collect your specimens.

Which Corals To Collect

You will need a face mask, snorkle, swim fins, and gloves as well as a small floating live well or collecting buckets. You will also need a poker, a piece of pipe flattened on one end, with which to pry up the coral. A pinch bar or large screwdriver will also do the job if you don't have a length of pipe.

As you swim over the reef, you will be confronted with a mass of coral. Nearly all of it will be unsuitable for your aquarium simply because it is too large. This is fortunate, for it means that the coral reef won't be carried home by aquarists should the live coral hobby increase a millionfold. The large corals will ensure continued propagation of the reefs. You must swim over the reef and look at the edge of it for the small corals. These may be growing in a clearing between the large heads or closer to shore in the gorgonian patches. Corals fist size or grapefruit size are best because they are easily handled and transported. A larger size, up to a foot across, may be occasionally collected, particularly of the moon coral, which will live well in the aquarium. I once kept a foot-high specimen alive for over a year in just an eight-gallon tank, so I am certain that in a large tank this species would do well with proper care.

Try a number of different species, including of course the moon coral, but try a good variety so you can see which species will do best. A small head of pillar coral (*Dendrogyra cylindrus*) six or eight inches

An elkhorn reef is a spectacular place to visit but no place to collect live corals or specimens. Elkhorn coral is generally too large for the hobbyist's or student's aquarium. Only a large public aquarium near the ocean could do justice in keeping it alive, for it requires strong current and ever-changing water. But if you wish to try it, pry off a very small piece from the edge of the reef. I photographed this beautiful reef about eighty miles off British Honduras.

The edge of the reef where the coral has broken down into rubble is the most productive. Here you can take small living rocks or small clumps of coral without damaging or destroying the beauty of the reef. Anyone who would find fault with this is an "eco" nut!

high, a six-inch slab of elkhorn coral (*Acropora palmata*), a section of star coral (*Monastrea annularis*), a cluster of finger coral (*Porites porites*), or the exquisite star coral, (*Eusmilia fastigiata*), which is found in small, fist-sized clumps, should all be tried. The leafy fire coral (*Millepora complanata*) is also worth trying. If possible select specimens that are inhabited by plume worms or fan worms, since this will be an added bonus and make the coral more attractive.

The brain corals (*Diploria clivoso*), low brain coral or the larger types (*Meandra*), are incredible to watch in the aquarium, for they change patterns right before your eyes. Small heads of nearly every type can be obtained, and sometimes the live coral tank does best if it is a little crowded rather than sparse, for the various species seem to aid one another in the confines of the aquarium. A large brain coral as a centerpiece in an aquarium can be a thing of wonder but it is not as hardy as the moon coral (*Montastrea cavernosa*), which in my opinion is one of the hardiest of all corals.

A huge gorgonian several feet across could conceivably be kept alive in a large aquarium if it were placed in a huge live well by itself and transported with great care, but this is a job only for the large aquariums and oceanariums. Those wishing to keep gorgonians alive should select finger-thick specimens less than a foot high. The small specimen at lower left behind the sea fan would be ideal.

The coral reef as it looks under the sea is always a breathtaking sight. But don't destroy it, for others will want to see it too.

Dense clusters of coral are best left alone. Do your picking at the outer edge of the reef where nature has already broken the reef.

Solenastrea hyades is a shallow-water coral found in five to seven feet of water in the Florida Keys. It is often found in a small head four to six inches across and may even be found in fist-sized heads lying loose on the bottom. It lives exceptionally well in the aquarium. A pygmy bass poses in front of the coral.

A small fist-sized chunk of living coral, *Solenastrea hyades,* growing upon an old conch in which a hermit crab has taken residence. Several plume worms also adorn the community.

Some of the shallow-water corals will do nearly as well, such as *Solenastrea hyades*, an inshore species, usually bright yellow in color. It is found in the Florida Keys in four to six feet of water, where it protrudes from the bottom in little heads six to eight inches high or larger. It lives very well. Mushroom or rose coral (*Manicina areolata*) is also quite hardy and is found in shallow water lying completely unattached.

There are several thousand species of corals and identification of them is sometimes difficult. Some corals may be the exact same species but will have a completely different form. Others will look the same but will be entirely different species. Common names mean little. Star coral is known by a dozen different scientific names, lettuce coral the same. Even the scientific names conflict at times, for there are few coral experts. I have prepared a large, well-illustrated book, *Exploring the Reef,* and have recently revised it, in an attempt to help stabilize at least the common names of many of the corals and have included the scientific names of many species basing them on the latest word. This is one of the few books to date that depicts most of the more common corals. Others are mentioned at the end of this book.

The corals are collected by prying them loose from the bottom with a small pinch bar or your poker. Insert the flat end of the bar under the base of the coral and give a sharp push downward or upward until the coral comes loose. The corals are firmly anchored to the sea floor and sometimes they won't come loose at all. If this happens, simply leave the piece alone and look for another specimen. You can usually remove the small heads more easily if you employ two pipes or rods. Lay one on the sea floor near the coral. Then insert the other under the coral, using one pipe as a fulcrum or lever. This will pop the coral loose quickly with little effort.

Care must be taken that the coral is not damaged when it is collected and the specimens should not be placed on top of each other in the collecting buckets. Set each piece upright in the bucket so they don't rub or touch each other. Keep them out of the hot sun and keep their container cool. Warm water will quickly kill them. A wet towel placed over their container will help to keep the water cool until you can return home with them. If you must keep them in a boat any length of time or in your car, insert an air stone from a battery operated air pump or dip up some water with your face mask or a plastic scoop and pour it back into the bucket to aerate it. A constant change of sea water in the holding bucket will be helpful while you are collecting more specimens or on your way back to the docks, should you be collecting from a boat.

Live Rocks

Live rocks may be found in shallow or deep water and these should be collected right along with the live corals, since they are every bit as interesting. A live rock is simply a piece of coral that has died but that has been taken over by a host of other marine organisms. It can contain a colony of tube worms, a rich growth of assorted algaes, numerous tunicates or sea squirts, burrowing clams, mollusks, live barnacles, and numerous live shells, brittle starfish, and hydroids. It can be a colony of life of great diversity. Naturally all live rocks are not the same. Some will have much more life than others and these should be selected. A single live rock a foot across can fill a small aquarium and be the center of attraction, making the aquarium complete. At one time while judging the saltwater aquarium show for the Aquarium Society of Broward County in Fort Lauderdale, I gave a high award, perhaps first prize, to a young man, Terry McGovern, who had set up

Two large "living rocks" transplanted to the aquarium. Though uninteresting to the beachcomber, a rock from an area rich in marine life can host a wealth of marine life, which can be studied at first hand with a magnifying glass. Such a rock should be transported, one to a bucket, with a small battery-operated air pump, the type used for live bait.

Incredible variety of marine life brought in on live rocks and maintained in the aquarium. This photo shows a solid mass of life including starfish, mollusks, nudibranch, anemones, and lesser forms. A blenny looks in from the lower right.

a spectacular aquarium with a single living rock and a few scallops. The water was absolutely sparkling clean and the many little algaes and growths on the rock made it look just like an undersea scene. No wonder it won. Another time, famed aquarist Tina Mann won the Best of Show at the society, with a natural tank of live rocks, barnacles, and algaes. They were arranged in a most attractive manner, as were all her tanks, and she kept the live tank in perfect shape for several years after the show.

When collecting live rocks or corals, make certain to pass up those which contain sponge, particularly the brown or black leatherlike sponge. This will usually die in the aquarium and may foul the water. Some of the small, erect sponges will live in the aquarium, but all sponges should be collected with caution because they are generally difficult to keep. Some of the inshore rocks will have a richer algae growth and a more diverse selection of plume worms, tunicates, and mollusks. These are found in just a few feet of water up to a depth of

This sponge crab has lost most of his sponge but it is still happy. It will reach up and take food from my fingers.

This cute sponge crab does a waltz with his "hat" which is really a hollowed out sponge that he carries for protection. If danger threatens it drops down the sponge, which appears as an inedible substance to the predator.

Live "rocks" will nearly always contain a sponge crab or two, and these should be collected along with the rock, since they are of great interest in the aquarium.

six or eight feet. Don't pick rocks with too much algae because much of it will die. Select specimens with a medium growth, algaes a few inches high, and let them grow in the aquarium. Also be sure to pick up the mollusks that are living on or under the rock, because these will add interest to your living rock. Any small crabs or shrimp should also be scooped up. Many will be living on or under the rocks and these too should be gathered up. Small hermit crabs and snails of all types should also be taken, because they are excellent scavengers in the aquarium. Small clumps of finger coral (*Porites*) will nearly always contain a number of small crabs and shrimp, including the sponge crab, decorator crab, cleaner shrimp, and mantis shrimp, as well as several species of brittle starfish. All of these make interesting additions to the aquarium.

Even small sea cucumbers are worthwhile, and they seem to benefit the aquarium by ingesting mud and sand as they feed. If you collect a live rock around a bed of finger coral you are likely to find all types of specimens in and around it, including not only the above, but many small interesting fishes as well. Sometimes you will catch small blennies and gobies right inside your rock specimen. If you wish to catch everything in and on the rock, place a large net under the rock before you lift it from the water. This will increase your haul.

Rocks from the outer reef, although not as densely covered with algaes, are still interesting and worthwhile. Select those with some growth on them and pick specimens with plume worms, mollusks, and bits of live coral. Sometimes bits of live coral will regrow on a piece of dead coral, and could in time grow into a large coral head, which is how reefs eventually reach the surface in the first place. The whole reef is built upon the dead and crumbled bodies of other corals. Northern aquarists may also find interesting live rocks on both coasts.

Marine Plants

Many marine plants or algaes will do nicely in the aquarium if healthy specimens are obtained. As mentioned earlier, marine plants are not at all difficult to keep. Some will even grow and fill the aquarium. I have never seen aquarium plants die and foul the water as reported by some authors, but perhaps the plants they kept were not suitable. Some of the algaes like the northern kelps would certainly foul an aquarium, so you just can't indiscriminately throw a bunch of plants into an aquarium and expect success.

But if you carefully select certain algaes, particularly the green algaes, you should have at least moderate success. An aquarium with beautiful green plants from the sea is a pretty sight. It is also a healthy

Select live rocks that are inhabited by plume worms if possible. This clump of fire coral houses several that presented me with a surprise. When disturbed, they emit a dense "smoke" from their top making them appear like an erupting volcano.

An aquarium with bright-green marine algaes is both interesting and decorative. Several species are shown here. A red cardinal fish hovers at the top. The green globe at lower right is *Valonia ventricosa.*

aquarium, for the algaes give off oxygen and absorb carbon dioxide. Small fishes will find a source of food by nibbling certain algaes, but larger fishes will consume the entire plant, so they should be kept out of the aquarium if plants are to be the main feature.

Some of the more popular tropical algaes are the caulerpas, especially *Caulerpa prolifera, C. crassifolia,* and *C. paspaloides.* All of them do well in the aquarium. They are found in shallow water, up to a depth of six or eight feet, and are on runners with little roots or "holdfasts" every few inches. The plant should be collected in its entirety, being careful not to break the main stem. Otherwise it may lose some of its vital juices. Reach down into the sand and pinch up a small portion of the body around each root cluster but try to avoid breaking the main stringer. Some of them, like *C. ashmeadii* or *C. paspaloides,* are on long stringers that may run two or three feet in length and a foot in height, so not many of these would be needed to fill an aquarium. *Caulerpa prolifera,* on the other hand, is rather small with leaves three or four inches tall and rather short runners. It is often found in brackish water near shore, under bridges, or under docks in boat basins, in Florida or the Caribbean.

The Merman's Shaving Brush *(Penicillus capitatus)* is a popular algae (most marine plants are classified as algaes) and is simple to collect. Look for a bright green specimen, then reach down into the sea floor and pinch up a teaspoon of sand or sea bottom with the plant to be certain you have obtained the root section. This may be rinsed lightly in the sea before placing in your bucket to remove excess sand or mud. *Udotea flabellum,* a tiny green sea fan, a few inches high, is also an excellent plant for the aquarium. It is collected the same way as Penicillus and is usually found in the same area. There are many other small, palm-tree-like plants in the shallows, some related to the Penicillus, and these are equally as good, Many are just two or three inches high when mature and look very much like miniature palm trees. They make a most attractive display when planted with a little thought and artistry.

Some of the calcareous algaes also do very well in the aquarium. Halimeda in particular, which is easily identified by its tiny, jointed sections, lives extremely well. Another type of probably the same species grows in large patches, several feet across. It is bright green but as soon as it is disturbed, it turns almost snow white. Leave it alone for a short while and it turns bright green again. This is an excellent algae for the aquarium and you can break off a chunk six inches or a foot across, since it comes apart rather easily. It may immediately turn white in the collecting bucket but it should get its green color when it has settled in the aquarium. It must be a good source of food, because

A simple collection of algaes including caulerpa and penicillus, finger coral, and an algae-covered rock give the aquarium an undersea look when arranged as they would appear on the sea floor.

Acetabularia is very common near shore oft tne Florida coast, especially in the bay areas. It grows upon loose rock and rubble. In this instance it has covered a discarded beer can in which a toadfish has made its home. The toadfish was guarding eggs that I hatched and reared to maturity.

many small angelfish live beneath it and can constantly be seen pecking at it on the reef as well as in the home aquarium.

Acetabularia is another unusual and interesting algae. It is sometimes called the Mermaid's Cup, for it has numerous tiny cups on the end of a three or four-inch stem. It is a delicate but beautiful algae, and if you collect it, you must protect it from fish or crabs that would otherwise quickly destroy it. It is easily gathered in very shallow water, often right from shore, where it is found attached to small rocks. The rocks and all are taken and kept upright in the collecting bucket as well as in the aquarium. This algae must have good light, preferably strong sunlight, so you may have to sun it in a bowl once in a while.

There are dense mats of rich green algae that appear to be a type of course hair algae, which can be collected from rocks or corals in the shallow areas. It can be found in masses a foot or two across and is bright green. If you start at one end, you can lift up the entire mat, much as you would a sod of lawn grass. The whole thing may then be placed in the collecting bucket. Do not place anything on top of it

Moss-covered rocks may contain tubularia, a colonial hydroid of interest to biologists in schools who rarely see a live specimen. In the aquarium the entire reproductive process may be studied.

because it may damage it. This is an excellent algae, and if you give it good light in the aquarium, it will live for years. I once sent a large clump of it to a doctor in Nashville, Tennessee, and a *year* later he called up and said that it was starting to fade and could I please send him another piece! I have often wondered why aquarists haven't used this algae more, especially at aquarium shows. If one were to completely carpet the bottom of a fifteen-gallon tank with it and insert one or two live rocks or corals with a few choice fish or invertebrates, they would have a stunning exhibit, surely worthy of a high award.

There are many other algaes that are desirable. The mossy algae or the sea moss (Bryopsis) as well as sea lettuce (Ulva) will do well in the aquarium and are easily obtained. The giant single-celled algae *Valonia ventricosa,* a bright green, grape-sized sphere, is of special interest because it is the largest single cell in the world. It is often found on the shallow grass beds attached to a small rock or coral. It will do well in the aquarium. Any of the small, bright green algaes can be collected, but stay away from the kelps or slimy alages, because these do not usually live well. If you want to try any of them, collect very small specimens, taking the rock or a portion of the sea floor with them, and put just a very few in the aquarium. It is always worthwhile to experiment. Otherwise we will never learn very much about the sea.

Most algaes will grow where there is a fairly strong current, especially the penicillus, udotea, and some of the caulerpas. This indicates that the aquarium in which they are kept should also have good current, which is probably how the plants receive their nourishment. It is carried to them by the current. The plants also need good light, so give them eight hours of light a day, or some sunlight if you can locate your aquarium near a window. If not, then use a Vita Light or fluorescent tube that simulates sunlight. Also you could carry the plants outside once or twice a month to sun them as you would the corals. Do not place too many plants in your collecting bucket at one time or they will die. Also keep them out of the hot sun, because warm water will quickly kill them. As soon as you get them home, plant them in the aquarium, pushing the root sections well down into the sand. Use natural coral sand or shell sand about two inches deep. Give good aeration and filtration, changing at least a few inches of water each month. As with the corals, it might benefit the plants to stir up the sand slightly to move dissolved materials about the aquarium. NEVER place any medication in with the plants, because this will quickly kill them.

If you wish to add fish or invertebrates to your live plant display, add only small fish such as very small angelfish, demoiselles, gobies, beanfish, cowfish, sea horses, clownfish, or pipefish. Do not add large

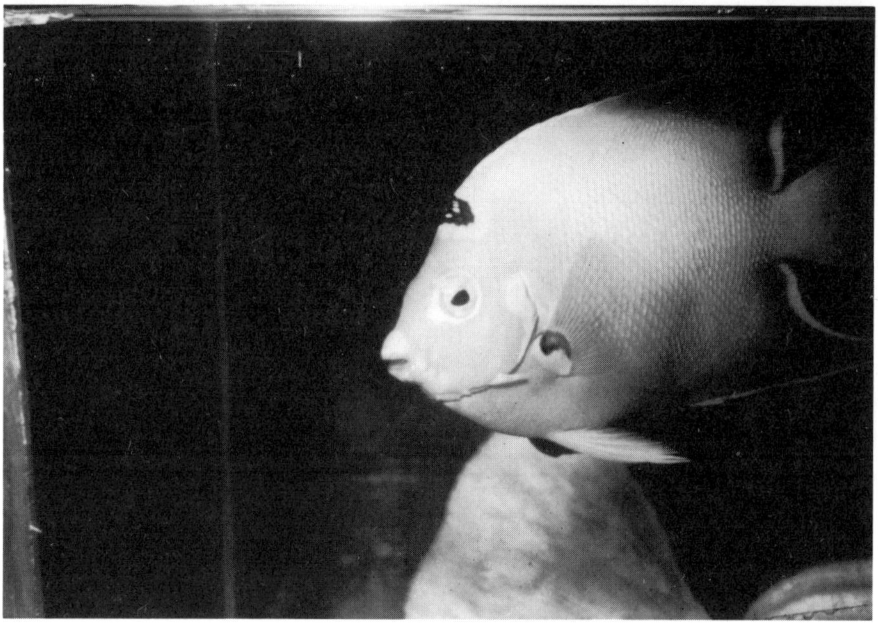

No aquarium can really be complete without at least a few fish. The townsend angelfish from Florida is said to be a hybird between the queen and blue angelfish. It will keep its color much better in a tank of live corals.

angelfish, tangs, or triggerfish or your plants will quickly disappear. Very small hermit crabs, serpent starfish, small sea cucumbers, coral shrimp, cleaner shrimp, and arrow crabs are also interesting in the live plant tank and will not do too much harm. Of course live rocks and live corals can be added, for the green plants will complement the corals, making a more complete undersea scene.

Eel grass or turtle grass can also be kept in the aquarium with varying degrees of success. It should be obtained in a sandy rather than a muddy area and the shorter blades three or four inches high should be selected. Simply dig up a healthy sod or clump, six to eight inches across. You may have to take a six-inch-deep lump of dirt with it to include the roots. If the roots go deeper than that, try another location. Choose a healthy section where the grass is rich green and cut out the desired portion with a spade, then dig under it and lift it out in one piece. It may then be placed in the aquarium on top of two inches of sand. Good aeration and filtration as well as strong light should keep it in good condition. It is generally not as successful as the other algaes, but it is a true plant and worth keeping if it can be kept alive. Sometimes pea-sized seeds of grass can be obtained floating near shore

The real beauty of a seashell is lost when the animal is killed simply to collect the shell. It is much more rewarding to keep it alive and no great task, for most seashells live very well in the aquarium, as does this marginella.

I brought in this coral from Puerto Rico and for a long while it appeared to be dead. But it looked good and had no odor so I kept it in the aquarium. One day it blossomed into this beautiful orange "flower," one of the many surprises I have had in keeping live corals.

Giant plume worms are especially fascinating in the aquarium. They live very well and will even reproduce if aquarium conditions are good.

One certainly could not study the deadly man-of-war (*Physalia*) close up like this in the open sea without danger of severe stinging, but in the salt water aquarium it is easily done.

or collected from a net drawn through the grass beds. These can be planted and they will grow if proper nutrients are provided in the aquarium sand. A little mud or dirt from the sea floor should help.

Miscellaneous Fish and Invertebrates

No book about saltwater aquariums could be complete without mentioning the exquisite coral fishes. These are the most colorful living creations of the world, true gems from the sea. They are rather easily collected with a small aquarium net a foot across. The net is made of plastic screen and is a foot deep. Or the sides of the net may be made of clear plastic with a screen bottom to let the water flow out.

All kinds of colorful fish can be collected with this little net. You don't need any poison or drugs as some people recommend, and I should know for I made my living collecting saltwater fish for nearly twenty years. You can catch every kind of fish you would want without knockout drugs, so don't ever let someone talk you into using drugs for collecting. It takes all the fun and sport out of catching a tiny fish. It is rather sad to see a drug collector squirting his poison around. The poor little fish floats lifeless in the water and he scoops them up,

A live coral aquarium containing multitudes of marine life as well as clownfish. Once the tank is established the fish will flourish in colors more beautiful than when kept in the sterile tank.

hoping they will recover. Fortunately the drugs are now outlawed in Florida for collecting fish, unless you get a special permit from the state, so forget about the drugs. You can catch all the fish you can use without them.

I personally prefer the hand net over the slurp gun for collecting, because the net is more versatile and you can catch any size fish, whereas with the slurp gun you can only catch fish that are small enough to fit into the barrel of the gun. All it takes to catch a fish with the hand net is practice and patience. The best way is to watch someone else do it, because this will save you a lot of effort, but if this isn't possible, then follow this brief lesson.

First, locate your specimen, preferably in shallow water where there is no current. Next, swim quietly over to it. It will usually swim under a rock or coral as you approach. Set your net over the opening in which the fish disappeared, then using the other hand, probe around or under the rock with a piece of pipe or a poker, watching the net to see if the fish goes into it. If it does, lift the net quickly to the surface and place your fish in the collecting bucket. If it doesn't, check the rock to see if there is an escape route or move away and let the fish move to a better spot.

Never collect in an area of honeycombed rocks, huge boulders or dense corals, because you will be wasting your time. Instead, collect in shallow, sponge, or finger-coral beds where there are small foot-wide corals and rocks. This way you will have a chance to isolate and net your specimens. Do not frighten the fish as you approach. Swim slowly, and once you have set your net in place, do not move the net, because this will frighten the fish and he will not swim into it. Move your hand or the poker to *drive* the fish into the net. Never try to swim after a fish and scoop him up. The slowest fish can swim faster than you and he will make a monkey out of you every time. Instead, set the net in place and *drive* the fish into it. It is easy once you have mastered the technique. Try some of the easy fish like baby beau gregories, high hats, angelfish, or cardinals at first. These are the easiest to catch. Also, go after the very small fish, because they are less alert. Don't bother with the sergeant major, because he is one of the quickest, most elusive little fish in the sea. Of course if he is in a tide pool at low tide and can't get out you may catch him eventually but he is mighty clever. Yellow tangs and butterflyfish are also more difficult to catch for the beginner.

You could compare catching a fish with a net to catching a butterfly on land. If the butterfly flew into an apple tree you would not be able to net it at all, for your net would become entangled in the branches. But if it landed in a meadow, you could easily slip the net over it. The

A prized find is this angular triton *(Cymatium femorale)*, which my daughter Julie Straughan collected at Big Pine Key Fishing Lodge in the Florida Keys while snorkling in the shallows there. It is considered uncommon, yet we collected two specimens in the eelgrass there. Rather than kill it as some collectors would do, we kept it alive in the aquarium.

same thing works under the sea. If the little fish is inside a mass of coral there is no way you can catch it, but if you get it out in the open, around a small rock or coral, you can easily slip the net over it to catch it.

You can also find many exotic fish around empty conch shells, especially in the Bahamas and Caribbean. Very often, they will swim inside the shell as you approach, so all you need to do is lift up the shell, holding your net under it. When you get to the surface, simply pour the water from the shell and the fish will come out. It's an easy way to collect and you can get a huge variety of fish this way with little effort. Down in the Florida Keys, you can also catch them this way in holes just inshore from the reefs. The conch shells are faded and white, but they provide a safe haven for the little fish.

Another rare find was this Atlantic hairy triton *(Cymatium pileare),* also considered a real prize by the collector. It was kept alive in the aquarium.

Underside of the hairy triton reveals beautiful coloring.

Live Seashells

Live seashells of all types are picked up by swimming over the grass flats with mask and snorkle. You will find them adhering to the blades of grass or crawling on the sea floor. On a recent trip to the Big Pine Key Fishing Lodge at Big Pine Key Florida, my daughter, Julie, and I collected the angular triton, hairy triton, murex, and live cone shells in a matter of hours. In fact if collecting is good, you can get all the fish and specimens you can use in a day or two.

The larger conchs such as the queen conch, horse conch, or the helmet shell, whelks, vase shells, pen shells, or tulip shells are all collected the same way. Simply swim over the grass or shallow gorgonian beds, preferably just before low tide, and watch for them. You can find practically every kind of shell this way, including beautiful marginellas, which are a prized shell, especially for the aquarium. These are easily spotted when they crawl, for they have a huge, brilliant

The queen conch (*Strombus gigas*) is common on the grass flats in the Florida Keys and Bahamas. It lives well in the aquarium. Similar shells like this Hawkwing conch are more suitable for the aquarium because of their smaller, fist-sized shell.

"Well, hello there!" this conch seems to be saying as it extends its long proboscis and stalked eyes. Most seashells live well in the invertebrate tank.

Immature conchs look much like cone shells and can fool even the experts.

mantle that is much more beautiful than the shiny shell itself. Its a shame that shell collectors nearly always kill the shells they collect. If only they knew how easy it is to keep live seashells. There is really nothing to keeping most of them alive in the aquarium. It's just a matter of setting up a few five- or ten-gallon tanks with sea water and sand. Then put in a few algae-covered rocks or plants and put in the shells. That is all there is to it. The shells will live for years. Some shells will lay eggs right in the aquarium and the eggs will hatch into little shells. It takes a little digging to find the diet of the seashells. Most are carnivorous and feed on other shells, so put a good variety of shells, clams, and snails in your tank so that those which feed on other shells will have something to eat. Many shells will feed on dead fish, clams, or algae. Most shell books tell something about the diet of the shells so that keeping of them is greatly simplified. One shell, the tun shell (*Tonna galea*), feeds on sea cucumbers as reported by Fran

The marginellas are small, prized gems in the shell world. The shell may be shiny orange but what a pity to kill it. Why not keep it alive in the aquarium as I have done here. This handsome specimen *(Prunum carneum)* was found crawling on a shallow grass flat at low tide. Its mantle is even more beautiful than the shell itself. It lives on a dense mat of green algae described earlier in this chapter.

Hutchings in her article in *Salt Water Aquarium Magazine,* Vol. 9 No. 2. Tulip shells feed on clams or other shells, so prized shells should not be kept with them. Some shells like the lovely angel wing shell (*Cyrtopleura costata*) live far down in the mud and stick their siphons nearly to the surface to draw in their food. They are handsome shells but extremely delicate to handle, especially when alive. I dug a specimen from a mud flat on Ceday Key while on a collecting trip with Warren Prince and my daughter, Julie, and brought it back alive. Despite the fact that part of the shell was crushed in wrenching it to the surface through the thick mud, it lived for several weeks in the aquarium. I am certain that a healthy specimen would survive indefinitely if some of the mud and water were taken from the area in which it was collected.

Shell clubs should alert their members to the idea of keeping their shells alive rather than killing them. They will find much more enjoyment, for very often the animal living within the shell is far more beautiful than the shell itself. Thus they will get double benefit from their shell and learn something about its habits as well. They might even be able to raise some of the shells in the aquarium and this would be well worthwhile, especially if the shell is very rare. They have everything to gain and nothing to lose if they keep the shells alive, for even if the shell eventually dies, they can still have the shell. They can easily pick the operculum out of the aquarium. A live shell tank need not be large. A three-gallon or five-gallon tank is fine for the smaller shells. Of course a large horse conch a foot long will need a big six-foot aquarium.

Cowries, those beautiful, highly polished shells, are not usually found on the grass flats except possibly at night. These are found under rocks or under coral heads. Carefully tip up the rock or coral slowly so as not to disturb the mud or sand beneath it and look up under the coral. The shells will usually be found clinging to the under surface or in a hole or depression there. You should also look on the sand under the coral as you lift it up, for some species will be hiding there. Fan the sand or probe it with your fingers to a depth of several inches. You may find some of the rare shells there. But be certain not to break or damage the coral and set it back carefully in place exactly as you found it so it will not die. Many creatures lay eggs on or under the coral, and if you expose the under surface of the coral to the sun without turning it back, these will needlessly die.

Brilliant flame scallops (*Lima scabra*), the file shells with the bright red interiors are found under coral heads or flat rocks. These can be pulled loose by grasping them firmly and are easily collected. They are among the most beautiful of scallops and a fine attraction in the

Like a flying saucer, a large moon shell emerges from the sand in the aquarium. This interesting specimen was collected from shore at low tide on Sanibel Island on Florida's west coast. Sanibel and Captiva Islands are world famous for shells. The aquarist can find live specimens and bring them back with a simple battery-operated air pump and a collecting bucket.

The moon shell lifts itself out of the sand and exposes its enormous body, which is nearly double the size of the shell itself. Note the tiny antennae in lower center of shell. Smaller bivalves are miniature coquina shells collected in the same area.

A live marginella with its brilliantly polished shell and delicate mantle can easily be kept alive. If it eventually dies, the shell may still be added to one's collection.

The common drill shell lives well in the aquarium but it will kill other shells by drilling a hole through their shell. It should not be kept with choice shells. The shell is shown laying eggs, which can be seen just above the left edge of the shell.

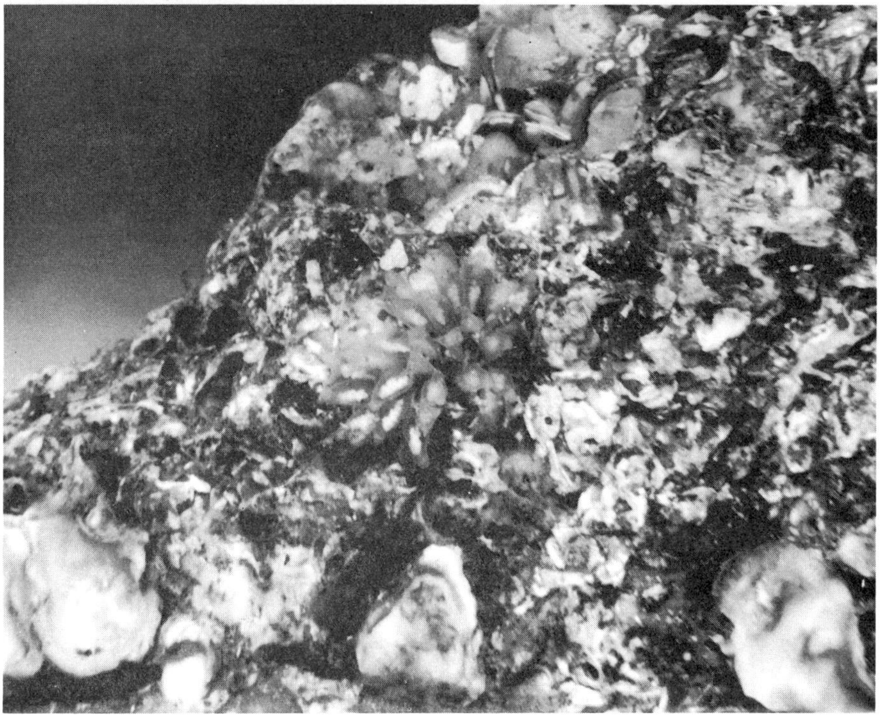

The egg case of the drill shell blends with the rock. Look in center of photo.

aquarium. Other scallops are equally interesting and many can be found in the grass flats on both coasts of Florida as well as in northern areas. The brilliant flamingo tongue shell with its orange spotted mantle is an aquarium favorite. It is collected on sea fans or gorgonians on the outer reef and is simply plucked off the gorgonian like so many grapes. You must learn to identify them with their mantle spread because this is how they will appear on the reef. Just look for bumps on the sea fans or gorgonians and you will see them. In the aquarium they must be provided with the same gorgonians on which you collect them if possible, because they feed on the living polyps. So when collecting them, pick a few gorgonians to keep alive and transplant these to the aquarium with the shells.

Red-and-white-banded coral shrimp (*Stenopis hispidus*) are found far under large coral heads. They are often found in mated pairs and if you catch a pair you may keep them together, but if you catch isolated specimens, keep them separate at all times or they will quickly declaw and kill one another. You can spot the coral shrimp under the coral by looking for its huge white antenna. Stick your poker under

A live horse conch *(Pleuroploca gigantea)* in one of my aquariums. This is the largest of Florida shells, reaching a length of over two feet. Only small specimens should be collected for the aquariums. Larger ones should be left in the sea, because they are becoming scarce due to dredging and mosquito spray in some areas.

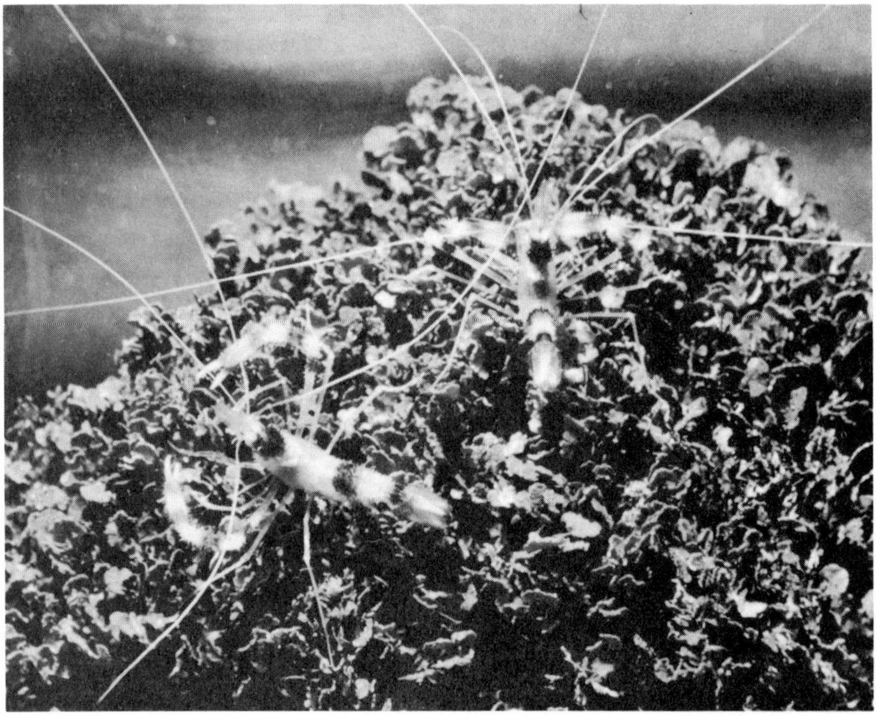

The banded coral shrimp *(Stenopis hispidus)* is the most colorful of all shrimp with its red and white markings and huge white antennae. It does well in the aquarium and will spawn. Raising the young, however, is very difficult. The green algae is a type of calcareous algae that grows in immense beds in the shallow water. When disturbed it turns almost snow white but when left alone, the bright-green color returns.

the coral, behind the shrimp, and you can easily guide it out to the front of the coral and into your net.

Red dancing shrimp or cleaning shrimp are also found under the corals, often very close to shore. Large rocks sometimes contain dozens or hundreds of them. Simply slip the net under the rock and scoop it back toward the top inside surface of the rock and you will catch them, sometimes a dozen at a time. The brilliant red, scarlet cleaner shrimp (*Hippolysmata grabhami*) is also found far under the coral, especially in deeper water. It is a prized specimen.

Live Gorgonians

Live gorgonians of all types are suitable for the aquarium. They are the "trees" of the reef and are classified as soft or horny corals. There are many different species. Most will do well in the aquarium if collected and transported properly. Simply grasp them firmly about the base and give a firm yank sideways. They will usually come loose. If they don't, simply leave them and try another specimen. Select specimens six inches to a foot high with just one or two branches and select the pencil-thick or finger-thick specimens, because these do best. Do not bother with the thin, purple, soft, slimy, and feathery types, because these do not do as well. Those with large polyps will be the best for the aquarium.

One species often found on the shallow gorgonian patches is alcyonium, called the dead man's fingers by biologists. It is purple in color and rather thick, and it often grows on dead gorgonians. It does exceedingly well in the aquarium and has huge polyps, making a most interesting display. Sometimes it can be found in clumps or lying loosely on the bottom. I have kept specimens in perfect condition well past the year mark with little effort.

Generally the single-stem gorgonians or small specimens six inches to a foot in length will do best. The larger specimens would need a very large aquarium and it would be difficult to transport them. But of course if you have a large aquarium, say six or eight feet long, it would certainly be worthwhile to keep one. All corals, rocks, gorgonians, or any specimens should be well aerated if you must transport them any distance. A battery aerator for each bucket with an airstone should keep them alive while you are transporting them. If you must travel overnight or several days, simply bring along a regular aquarium air pump as well and hook this up at night to furnish aeration while you are resting.

The biology student with special interest in small creatures could obtain rocks with colonial hydrozoans, small fernlike creatures that

The common star shell is plentiful on the grass flats in the Florida Keys and Bahamas. It is easily kept in the aquarium. A shell tank need not be large, and two-gallon tanks are sufficient for the smaller shells. Shells can be collected from Maine to Florida on to the West Coast. However, California has a law that makes a child a criminal if he picks even a shell. This law should be repealed, for it is unwarranted. A small group of researchers there apparently are trying to keep the state's gigantic coastline all to themselves.

grow on dead corals or gorgonians. These can sting like fire coral so must be handled carefully. Always wear heavy, neoprene rubber gloves when collecting to protect the hands from cuts and stings. I wear the heavy type that electricians use. They come up to my elbows and give perfect protection, not only to my hands but to the arms as well. Rocks from the ocean in more northern areas may contain minute tubularia and the strange skeleton shrimp, caprella, or the little sea spider, nymphon, which is all legs and no body, a strange little creature. A single rock from a rich shore area will contain a fantastic wealth of minute life that can provide weeks or months of interest and discovery for the student with a magnifying glass or low-power stereomicroscope.

The little sea spider, nymphon, is of special interest to the biology student. It is all legs with no body cavity at all. A moss-covered rock in an area of rich marine growth will yield specimens, but it takes a trained eye to see one.

Nymphon, the sea spider "ain't got no body." This fine specimen lived in one of my algae rock aquariums.

There is no limit to what can be found in the sea, and fortunately practically all of it can be kept alive in the aquarium for varying lengths of time. The life factor of many species will increase as we learn more about them and in time many species will actually be raised right in the aquarium. This has already happened to a certain degree.

Collecting Laws

Some states have laws regarding the collecting of marine life. Unfortunately California has banned *all* tide-pool collecting throughout the entire state at this writing, unless I have been misinformed. Oregon has similar laws. This is unfortunate and unwise and it should be corrected. What harm can a youngster do with a bucket and a little hand net? Instead of being at the seashore he could be at home getting into trouble. Picking a few critters and rocks from the sea is a clean, healthy avocation. It should be encouraged, not discouraged. With thousands of miles of shoreline, the aquarium hobbyist isn't going to destroy the ocean! Whoever passed this law was badly informed and the law should be repealed at once. To make it a crime to pick up a live sea urchin or hermit crab is a disgrace to a free America when there is so much real crime in our country. To destroy someone's interest in the sea is a tragedy and this is what is being done. It should be stopped. Those living in this area can get this law changed by writing to their governor, local newspapers, and congressmen. If some authorities want to protect a certain area from everyone, then let special little undersea parks be set aside, but not the entire coastline. This is an outrage. Let the aquarium clubs, shell clubs, biology and science students all speak up. They are voters too, and they can change this unfair legislation.

Florida also has some collecting laws but fortunately they are much more sensible and fairer than California (at least at present). There are now laws protecting certain species of coral such as elkhorn coral (*Acropora palmata*), staghorn coral (*Acropora cervicornis*), and pillar coral (*Dendrogyra cylindrus*). These corals are now protected in Florida and you are not allowed to take more than ten specimens or ten pounds of them. Sea fans are included on the restricted list. The ten pounds is your *total* weight of all specimens combined, not ten pounds of each. I feel that this is a sensible and needed law, for these corals are in short supply. In Florida waters they could easily be stripped off the reefs in time if some controls were not placed on them. There are plenty of other corals that are not protected (I don't believe the law covers live rocks) and the new law still allows you to take at least some of the protected species for souvenirs or study (at

On a coral reef in the Florida Keys I am shown photographing some of the beautiful coral formations. Coral is extremely sharp and should only be approached closely by an experienced diver. Beautiful formations such as this should never be disturbed by the student or collector, for their beauty is a priceless heritage. Collecting should be done on inshore rock patches or near dead reefs. *Courtesy James La Tourrette.*

A youthful collector, my daughter, Julie, drags a small net through the eel grass in the Florida Keys in search of live seashells. Many fine specimens may be collected this way.

this writing *all* corals may come under the ten-pound limit). Conchs are also protected. I think you are allowed ten per day but it is best to check with the conservation department or local diving area for the latest rules and regulations. The place where you stay or rent your boat should be able to inform you of all the local rules. No collecting of corals of any type is allowed in Pennekamp, because this is a coral preserve.

If you don't live by the sea, you will either have to buy all of your specimens or you can set up your aquarium and then go to the tropics for your vacation to collect your own. Bring a few Styrofoam boxes or at least one or two and a number of plastic bags. Use triple plastic bags for live rocks or corals and fill the bag half full with clean sea water so that it well covers the corals. Then fill up the bag with air or pure oxygen if you can get it and twist the top tightly, bend it over, and seal it with strong rubber bands. As you travel, the water will slosh

If you venture to the seashore on a collecting trip, a couple of styrofoam boxes with large plastic bags and large collecting buckets will allow you to bring back live coral and shells in perfect shape. A regular air pump is used in the motel but when you travel, switch to the small, battery-operated pump. Plastic milk jugs will hold extra sea water for the trip home.

Gobies and blennies are very common in the tide pools and you may catch them as you lift rocks from the sea floor by simply holding a net under the rock. They are extremely hardy and perfect for the invertebrate tank.

The weird batfish is also found near shore in brackish areas frequented by the invertebrate collector. It is easily collected by placing a net over it or grabbing it from behind. Despite its fierce appearance, it is totally harmless. It feeds on baby hermit crabs and small minnows.

back and forth in the box and keep the specimens alive. Don't try to bring back too many specimens in one box or you will lose them. If you are driving, put your smaller specimens in the boxes and put one or two chunks of live coral in a large bucket. Then insert an air stone from a battery-operated air pump and let it run for the duration of your trip. Should you need to stop overnight as mentioned above, bring along a regular aquarium pump and a gang valve and bring your specimens into the motel where you can aerate them.

Make certain that you check the area in which you plan to vacation to see if they allow collecting of specimens. Some areas like the Bahamas *do not* at present, so you should spend your money in an area that will let you enjoy your hobby. How a country like the Bahamas, with thousands of square miles of reef, could feel that a hobbyist with a hand net and a face mask is going to make a dent in the fish population or destroy the reef by picking up a chunk of coral is beyond me. They will allow a resort area to dredge and fill in an entire shore and reef area, or let a mining company dredge up sand by the thousands of tons. This of course kills an enormous amount of coral and marine life. But perhaps with their new independence they will change this archaic ruling and welcome the skin-diving tourist who simply goes elsewhere at present.

Keeping the ocean alive in your home is a rewarding hobby, literally without end, for the ocean is almost endless and we have just begun to explore the great depths of the sea. Even if your reef should die, you can still save it forever, for you can remove the coral or corals from the tank and wash them under a strong jet from the garden hose until they turn white. Then place them in a bucket of water with a half cup of bleach for a few hours. Rinse them again in fresh water and let them dry in the sun. You will still have your corals and they will be even more beautiful than ever. Furthermore, they will keep forever and maintain their beauty as long as you live. Now what other hobby can offer that?

6

Hints on Care and Feeding

IF YOU KEEP FISH IN WITH YOUR CORALS AND LIVE ROCKS, THESE WILL require special foods, depending on the species. Angelfish, Pacific batfish, and clownfish will eat adult frozen brine shrimp, Tetramin dry flakes or other dry foods, chopped shrimp or scallop, and bits of lean beef heart. Butterflyfish will need newly hatched live brine shrimp, very fine shredded scallop, and live tubifex worms for some species like the four-eyed butterfly (*Chaetodon capistratus*) or the common butterflyfish (*Chaetodon ocellatus*) and other fussy feeders. If the live worms cannot be obtained, try them on frozen tubifex or freeze dried. Soak the dried worms or the frozen ones in fresh water until well thawed or well soaked before feeding.

Larger fish like the triggerfish, grouper, or snappers will eat chunks of fresh shrimp (quarter-inch to half-inch chunks), small minnows, or fresh clam or scallop. They will also eat many of your small crabs and shrimp, such as the arrow crabs, coral shrimp, cleaner shrimp, and small hermit crabs, so if you value these specimens, keep the big-mouthed fish in tanks to themselves.

Adult frozen brine shrimp is an excellent food if it is obtained in good condition. But if it has been thawed and frozen, it may be totally worthless. Make certain that the shrimp are pink and firm and have never been thawed when you purchase them. I often mix a little dry food in with the shrimp when feeding to give the fish and corals a variety. I break off a chunk of the frozen shrimp and place it in a glass of fresh water to thaw. Then I mix in a good pinch of the flake food, usually Tetramin, which I stir gently into the water. I place the food in the aquarium with a small plastic spoon. If there is any food left over, this is kept in the refrigerator in the glass of water until it is needed. The water keeps it from dehydrating.

Fresh eating shrimp is also an excellent food. Never use frozen bait shrimp, because these are often contaminated. Also they may contain preservatives that will be harmful to the fish. Live bait shrimp are alright, but these should be frozen to kill harmful parasites or copepods. Regular eating fish is good if fed sparingly.

The diet of marine fish varies greatly, for many fish have specialized diets. My book *The Salt Water Aquarium in the Home,* which contains nearly four hundred pages, covers practically all of the commonly available fish as well as many of the invertebrates, and it would be a repetition to list all of them here.

Baby marine fish, especially tiny, half-inch angelfish, triggerfish, or butterflyfish, will usually thrive in the live coral tank, for they will find good nibbling on the many corals and algaes. I have raised many baby fish in flawless condition, up to a marketable size in my coral tanks. Some would be delicate Pacific specimens of great value that surely would have died in a regular aquarium. The newly hatched brine shrimp and fine baby dry food, fed to the corals, also fed them and they thrived.

The spectacular lionfish *(Peterois volitans)* living in an aquarium of live corals. The specimens are in perfect condition with no signs of disease or parasites and were raised with dozens of others in the same tank from tiny two-inch babies. Note beautiful moon coral to the left.

The aquarium of snow-white corals will eventually become obsolete. This tank is being transformed to an invertebrate tank. Simply remove some of the corals as you obtain live rocks. The fish and specimens can remain undisturbed.

The same tank with all white corals removed. Thus the tank was transferred from a sterile tank to an invertebrate tank. No medication of any kind was ever used, yet the fish and specimens flourished.

I have often stated that live corals and live rocks greatly reduce or totally eliminate most diseases in the aquarium and I am more convinced of this than ever, having run numerous tests to evaluate my findings. Aquariums that contain lethal parasites that would kill any fish placed in the tank in a matter of days or hours, would become completely safe when that same tank was filled with live corals. At this point, I am not certain exactly what does it, whether it is the anemones, the coral polyps, or the rocks, but I am convinced beyond all shadows of a doubt that it works, having tried it many times for over twenty years. I suspect that the numerous corals, mollusks, and gorgonians filter out the harmful microorganisms from the water or that they are kept under control by microorganisms living on and within the coral. But at any rate, the live corals and rocks positively will rid the aquarium of most disease and parasites in short order. This is also substantiated by aquarist Lee Chin Eng of Indonesia, who first wrote to me around 1962 and who is a great advocate of live rocks

Healthy black angelfish living in the same tank with no water change. Gorgonians and live rocks seem to remove harmful organisms from the water, or perhaps they add an essential substance to the water, which is important to the survival of the fish.

and corals in the aquarium, stating that the live coral tank makes the bleached coral tank obsolete. I agree. But at that time, Mr. Eng did not use a sub-sand filter and according to his letters and pictures, he kept mostly live rocks rather than corals, apparently placing them on end rather than flat in the aquarium to avoid build-up of gases. Of course I nearly always recommend the undergravel filter, the only exception being if one is raising saltwater fish, for the newly hatched fish could become trapped in the gravel. Corals placed in the aquarium without the filter soon develop white mold or fungus areas and will die unless removed from the aquarium. But still I would not doubt Mr. Eng's success, for he is without doubt an astute aquarist and his wise and careful selection of suitable rocks and algaes has led to his success.

It would take dozens of full-length encyclopedias to tell everything about every aspect of the saltwater aquarium field, for we are dealing with the sea, which in itself is extremely complex. This book is intended to give you merely a start; an idea of how to go about keeping a segment of the coral reef and the undersea world alive in your aquarium. It will help if you already have aquarium experience, for you will then know about filters, heaters, and other aquarium apparatus. Your local aquarium dealer will be able to help you out in many situations, and if you are his customer, he will be happy to answer your questions.

A few things to remember are: don't start out too big. A ten- or twenty-gallon tank is fine. Five ten-gallon tanks are better than one fifty-gallon tank, especially for the beginner. Don't spray around the aquarium. Most insect or paint sprays are deadly to fish or specimens. Keep only very tiny fish with your live coral tank. Large fish will eat the plants and corals.

There is much more to remember. If you turn off your outside filter for longer than a day, drain all the water from the filter. Then rinse it in fresh water before use. If you let the water stand in it, it could cause deadly hydrogen sulfide gas to form, which may harm your specimens when this water is pumped into the aquarium. Never use fresh carbon without first washing and soaking it. Place your hands in the aquarium as little as possible. Use an aquarium scraper to move things about instead of your hands. I could go on and on, like a mother first sending her child off to school.

But you will learn and you will be fascinated. Don't worry too much about all the little crabs, anemones, and invertebrates. They will get their food from the bits of food you feed the live corals. But if you have a special pet like an arrow carb or a brittle starfish, hand it a chunk of fresh shrimp on the sharpened end of a wooden dowl. It will quickly learn to take food from you. One of the advantages of the live tank is that there is a surplus of foods for the little fellows. Baby angelfish, baby hermits, snails, and all the wonderful little creatures that make up your "ocean" will find plenty of food on their own. In fact if you didn't feed them for a week, they would still be around fat and saucy and probably a little larger.

If you have a saltwater aquarium now and it has been nothing but trouble, switch to a live tank. First, remove most of your corals, clean up the bottom of the tank with a fine filter net, then change about half of the water. Filter the aquarium with well-aged activated carbon and put in an anemone and some live rocks or coral. Go slowly at first in case your water is poisoned from all the copper and medications you may have put in it. But if the anemone and algae rocks live alright, you can add more, perhaps changing a little more of the water. Soon you will have a living reef and your fish will be healthier than you ever thought possible. But keep a quarantine tank on hand and if your fish need treatment put them in it. Never add any more medication to your main tank, for you won't need it. It will take care of itself and you will find the marine hobby a joy once again.

Nearly all of the invertebrates are absurdly simple to keep if you set up the aquarium properly at the beginning and if you start out with fresh sea water or at least a little of it in your artificial sea water mix. Snails, crabs, and starfish in particular can withstand almost total neglect and abuse and yet live on well past the year mark. About the only thing you would need to keep them indefinitely would be to keep the salinity down by adding fresh water every month or so.

I have kept starfish, sea anemones, live seashells, and all types of minute sea growth alive and in good condition with no more attention to the aquarium than an occasional feeding of *fine* dry food or fine chopped shrimp. Anyone who has difficulty in keeping anemones or starfish alive either has obtained bad specimens at the start, is using improper sub-sand filter material, has bad aquarium water, or poor quality sea salts. I use only reef salts in my work but other salts are good. Beware of absurd claims, for the salts are much the same.

Saltwater fish may at times be a problem but invertebrates are a breeze. Not only will they live on and on but they will even reproduce and grow up in the aquarium, even in the neglected aquarium. It appears that too much fuss and worry with the aquarium can do more harm than good. I always recommend the very simplest and basic of aquarium equipment and it nearly always is the most successful. I have had seashells reproduce and grow right up to half-inch size in old aquariums. I have also had starfish lay eggs and grow up in the aquarium. But recently, I was astounded to find a three-inch-high sea

If your saltwater aquarium has proven difficult, switch to the invertebrate tank. It's a breeze! Beautiful anemones such as this will thrive with little or no effort on your part.

pen growing in one of my aquariums. I had never collected it. In fact I had placed nothing in the aquarium except a couple of moss-covered live rocks and a few sea spiders and crabs when I set it up well over a year ago. The tank is run with a simple sub-sand filter and silica sand mixed with natural shell sand. There is a small inside bottom filter with a handful of old carbon. This is all the care and filtration the tank has received since it was set up. It has never been cleaned and yet all kinds of things are living in it, including anemones, crabs, and snails. A small stone crab, less than a quarter inch when collected, is now well past the inch and a half mark. But the sea pen was a pure surprise for I have no idea how it got into the aquarium. It must have been born there.

I would like to emphasize that the invertebrate aquarium, or even the saltwater aquarium with a full complement of colorful fish, does not need a lot of costly and elaborate equipment to be successful. The expensive gadgets like the power filters, ozonizers, protein skimmers, ultraviolet filters, and all the rest are often sold to the unsuspecting

Baby starfish born and raised in the aquarium. These are living in the filter where they were picked up by the siphon tube.

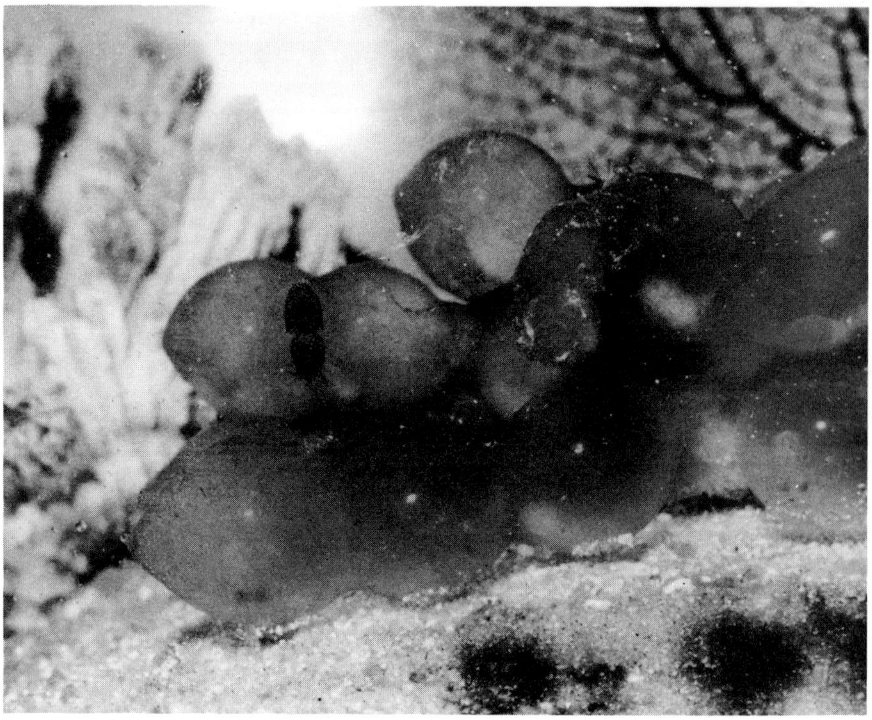

One of the rewards of the natural reef aquarium is the surprise and rewards it will bring. The author collected the above eggs and was informed that they were eggs of a sea shell. Imagine the surprise when they hatched out into little squids. This was the first aquarium hatching of squids ever recorded.

aquarist, especially the beginner, when in reality *none* of them are needed. I have kept many successful aquariums with nothing but the undergravel filter and a small air-driven outside filter with carbon. I have also seen hundreds of aquarists do the same. It isn't that the expensive gadgets are no good, for sometimes they can be useful for a specific problem. The power filter would be useful for a large aquarium, the ultraviolet filter would be useful to control disease and bacteria without having to medicate the tank. But all in all, the aquarist can dispense with most or all of the gadgetry and still have absolute success. Even nitrite test kits are not essential, although some aquarists find them useful.

Aquarists kept sea anemones and other invertebrates alive more than a hundred years ago and often without even an air pump, so the aquarist today surely should be able to keep them with a minimum of elaborate and expensive equipment. Live rocks seem to help a great deal. They furnish vital substance to the water, conditioning it and making it healthful to the specimens. In addition, they provide basic food for the crabs and shrimp when the aquarist neglects to feed them. I have kept aquariums without any aeration just to see how they would do and have kept crabs, live shells, and anemones for months on end with no air at all. Once in a while I would blow through the bubbler stem to circulate the water. One tank with live plants grew even better this way than when the tank was operated with an air pump. Of course I don't recommend keeping a tank without an air pump since the pumps are so inexpensive. I just tried it as an experiment. If sea anemones, hermit crabs, live snails, and starfish will live with no care and without even an air pump, then the aquarist should be able to keep them with little effort, especially when the tank is filtered and aerated.

The biggest problem in keeping invertebrates is usually from predation. In the ocean, everything eats everything else. The big ones eat the little ones, the starfish eat the clams, the shells eat the starfish, the crabs eat everything, the fish eat the crabs, and the aquarist eats the lobsters when they get big enough to eat! Consequently, if you wish to keep a special invertebrate like the coral shrimp, rose petal shrimp, or scarlet cleaner shrimp, be mighty careful what you keep with it or it may well be eaten.

Never keep large crabs with any smaller, delicate shrimp or they will quickly kill and devour the shrimp. Even hermit crabs can be killers. I once had a large hermit crab that would play "possum" in the aquarium. It would emerge part way out of its shell and pretend to be dead or asleep. When a fish moved within range it would quickly wake up and grab the fish with its powerful claws. I had placed the large hermit on exhibit in Aquarama, a large traveling aquarium that I had set up with fish and specimens when it first made its debut, and the owners, Ed and Sylvia Morehouse, had wondered what was happening to the fish in one aquarium. Careful observation proved that the large hermit was the culprit. Another time I saw three Florida spiny lobsters gang up on a sea turtle and try to drown it. They would all approach the turtle at the same time and grab it simultaneously, pulling it to the bottom and holding it there in an attempt to drown it. If we hadn't intervened and moved the turtle to another tank, they would certainly have drowned it.

Certain types of sea anemones will catch and kill all types of crabs, shrimp, and fish. I once saw a common Florida pink anemone (*Condylactis passiflora*) kill and devour a six-inch adult sea horse. Anemones can sting, and although some small crabs and shrimp are immune to their stinging cells and can crawl across them with impunity, other crabs and shrimp cannot and will be killed if they wander into the tentacles. Generally the small, half-inch anemones won't hurt much and you can keep rocks with dozens or hundreds of them in your aquarium with little worry, but the large anemones, especially the clown anemones (stoichactus), can be killers to both fish and invertebrates, including other anemones.

The same is true of crabs, shrimp, lobsters, and shells. All will feed on the others when and if they get the opportunity. A large lobster, whether it is the spiny lobster or the northern lobster, will eat anything that comes within reach as will any large crab, especially the blue crab. This is a killer crab if there ever was one. We used to pull off their main claws when we placed them on exhibit (and ate them) or they would quickly decapitate each other. They would grow new claws quite rapidly and when the new claws were large enough, we would again remove the claws and place them in the cooking pot.

The invertebrate tank need not be large. I have kept many interesting specimens in two-gallon tanks, like the fiddler crabs shown here. Since most marine creatures prey upon each other, it is often more practical to isolate the specimens in small tanks like this two-gallon plastic tank. A sub-sand filter keeps the water fresh and clear.

A living colony of coral nearly a foot across was transferred to my aquarium in a single piece. It contained live coral polyps and many tube worms. Brown material to the right is sponge growth and should be avoided, since it does not do well in the aquarium. A beautiful clownfish in perfect health enjoys this natural habitat as does the large sponge crab on the left.

An aquarium with anemones, live plants, and the purple alcyonum is easily kept in perfect condition with proper light and filtration.

A small live coral head showing multicolored fan worms. It is a simple matter to pick a live coral head and bring it back alive for close study and observation in the aquarium. Much more can be learned about it than by observing it under the sea. Periodic changes of fresh sea water and good filtration will keep it alive for a long while.

The flame scallop (*Lima scabra*), with its brilliant red interior, is the most interesting and colorful of the scallops. I was the first to collect them for the aquarium trade and gave them their common name.

Many of the crabs and shrimp can hold their own in the community aquarium while their skin is hard, but once they shed their skin, their body is as soft as jello and anything can and will eat them. I once saw a tiny rock skipper suck down a three- or four-inch arrow crab, like a piece of spaghetti, when the crab had shed its skin. It approached the crab and sucked it down from its needle nose in one quick slurp! If you wish to keep a special invertebrate like a mated pair of coral shrimp and hope to keep them for a long while, keep them in an aquarium to themselves with a small piece of live rock with a ledge or overhang so they can get out of the light. The tank need not be large. A three- to five-gallon tank is sufficient. I have kept many of them in gallon aquariums with complete success. In over twenty years of marine fish keeping, I have found no evidence that a large aquarium is better and more successful than a small tank, although one hears this claim quite often in the marine field. I have seen many very successful two- to five-gallon tanks. I have kept as many as twenty-five small fish and specimens in a two-gallon plastic tank and even had the fish spawn repeatedly, so I never advocate a large tank in preference to a small tank unless one is keeping very large specimens.

A battery of small tanks each with its own particular species of crabs and shrimp is certainly more practical than placing them all in one large tank, for you will have better control with the smaller tanks and maintaining them will be much less work. A sheet of glass or plastic on the top will keep evaporation at a minimum and any small inside or outside air-driven filter with a little carbon will adequately filter the tank. The sub-sand filter alone will keep most small tanks in perfect condition without any other filter if you will just replace a fourth of the water each month. Also, you will find small carbon cartridges that will come with some of the undergravel filters. These will keep the water from discoloring under normal circumstances. They should be rinsed well before use.

Keep in mind that all invertebrates feed upon each other. The tun shell eats the sea cucumber, the starfish eats the mollusks, but some mollusks like the triton will in turn eat the starfish. Many shells will feed upon other shells so you must learn which are carnivorous and don't keep prized shells with them. This is why I stress the small tanks so you can divide up your specimens rather than have them devour each other. Of course if you collect your own, this is not as serious a problem. You can then set up a fairly large tank with live rocks and corals and fill it with all kinds of sea shells, starfish, and crabs and let everything feed on each other as they do in the ocean. There is nothing wrong with this, for it is merely a matter of survival and you may discover new facts about marine life in general, particularly in regards to feeding. You can supplement the feeding of the various organisms by sprinkling a little dry, baby fish food or regular fish food on the surface of the tank each morning and night. It will eventually settle to the bottom when it gets over the bubbler stem and many of the various creatures from snails to crabs and shrimps will eat it. The starfish will also pick it up from the bottom as will the sea urchins.

There are many different types of invertebrates and the keeping of them may vary, but in general a large algae-covered rock will add food and minerals to the diet and this may be supplemented by sprinkling a little dry food on the surface as explained above. The baby dry food is good for small anemones, live corals, tubularia, and other minute growths, but this should be augmented with daily feedings of live, newly hatched brine shrimp, fine chopped shrimp, and other foods.

7
Twenty-five Popular Invertebrates

Banded Coral Shrimp *(Stenopus hispidus)*

THE BANDED CORAL SHRIMP IS PROBABLY THE MOST POPULAR OF ALL the crabs and shrimp and is readily available at aquarium stores handling marine specimens. It lives in the Caribbean, Bahamas, and Florida and is found all throughout the tropics not only in the Atlantic but in the Pacific as far as Australia. The most common is the red and white variety, and with its huge white antenna it is one of the most spectacular and colorful members of the shrimp family.

I have personally collected hundreds, perhaps thousands, of them alive for aquariums and consequently have learned a good deal about them. They live on the underside of the coral usually far back under a coral head or ledge. Quite often they are collected in mated pairs, and if a pair is collected together, they should be kept in a container to themselves by the collector and not placed with other shrimp. If just one shrimp is collected, it should be placed in a container by itself or with fish but not with another coral shrimp. If two shrimp are placed together and they are not mated, the stronger will usually quickly declaw and perhaps kill the weaker specimen. Without their brightly marked claws, they are not nearly as spectacular, and the astute collector is aware of this, taking great pains to seperate the shrimp. When collecting coral shrimp, I usually place either the mated pair or a single specimen in a gallon milk jug that has a large hole cut in the side. Once the shrimp are netted, they must be handled ever so gently or they will throw off their claws by themselves in an effort to escape. It takes a little practice to handle them so they won't declaw themselves before you can place them in a bucket, and sometimes even with the best of handling, they will throw off their claws at the last minute. But all is not lost, for they will grow a new set of claws the

A coral shrimp in its normal upside-down position in one of my invertebrate tanks. A sea cucumber crawls beneath it.

first time they shed their skin in the aquarium. Usually when I collect them, if they throw off their claws, I keep them in the aquarium until they grow a new set before shipping them out.

The coral shrimp is sometimes called the coral crab or candy-striped shrimp. It is a cleaner shrimp and will groom moray eels and other fish when they approach it for grooming. It apparently removes fungus or parasites from their bodies. Rarely will a fish eat the coral shrimp when it has offered to groom them. The fish will swim up to the home of the coral shrimp and then swim toward the shrimp with a sideway movement of the body, a signal that it would like to be groomed. It will slowly edge its body up to the shrimp, who will then climb aboard and groom the fish. It's a special service of the shrimp to its fellow creatures under the sea. Other shrimps also groom the fish and a number of small fishes like the neon goby, labroides, spanish hogfish, and others also groom the larger fish, removing fungus and parasites from their bodies.

There is a smaller subspecies of coral shrimp that is yellow and red. It lives in the same area as its larger relative and sometimes the two types may be found under the same ledge but never paired together, at least to my knowledge. The yellow variety or subspecies (whichever it is) is about half the size of the red and white species, giving it about a maximum three-inch length compared to six or seven inches for the red and white species.

Both are very hardy in the aquarium. They will eat dry food, small chunks of fresh fish or shrimp, as well as algae from rocks placed in the aquarium. The female is usually much larger than the male and at certain times will be loaded with eggs that will form a green mass on her body. The eggs are carried under the tail and will hatch out in the aquarium. I have never made a serious attempt to rear them, although one time I did rear some through the second stage. It is not an easy task, however, for the rearing of any marine shrimp is rather complicated, requiring special algae and culture of minute marine organisms upon which the larval shrimp must feed. They go through several stages before taking the form of the adult shrimp, and it is the rearing of them through each of these stages that is difficult. But it will come in time, for already commercial mariculture farms are raising vast quantities of eating shrimp, which is no more of a task than rearing the coral shrimp. Probably at the time this book goes to press someone will be raising and marketing the banded coral shrimp and other species for aquarists. Already the coral fishes themselves have been spawned and reared in quantities of thousands by marine biologist Martin Moe as reported in *Salt Water Aquarium Magazine*, Vol. 9, No. 2, the world's first marine aquarium magazine. (I have published and edited it for nine years.)

Coral shrimp can be kept in the community aquarium with fish, but make certain there is plenty of coral rock under which they can hide. Prop the rock so that there is a good opening beneath it because the shrimp like to crawl upside down on the under surface of a rock, and keep the light a little on the dim side. Don't keep large-mouthed fish like groupers, snappers, or triggerfish with the coral shrimp. They may not bother their little "cleaner" friend, but there is no way to be certain.

The Arrow Crab

The arrow crab is sometimes called the spider crab or "old needle nose," and it is a long-time aquarium favorite. Like the coral shrimp, it is best kept just one to the aquarium unless one has obtained a mated pair. Small specimens may be placed in the same tank if there is plenty of room, but large specimens are likely to fight. They will not only declaw each other but will pull off each other's legs until all

The arrow crab is a favorite in the aquarium. It uses its long legs to probe or ward off an intruder. Its needle nose is used as a larder. Surplus food is attached to the hooks there.

An arrow crab in the invertebrate tank will usually find some food on the algae rocks and corals, but it should be handed a small chunk of fish or shrimp on the end of a wooden dowl. It will quickly accept the handout.

that is left is the body itself which will still remain very much alive, although a rather pitiful sight. I am not certain, but I feel that if the body is removed to a protected tank it will regenerate all of its lost legs. It seems that I did this once or twice.

The small spider crabs up to two inches are less likely to fight. They are also more colorful than the adults and are a treat to watch in the aquarium. They will preen themselves, taking great care to clean each long leg as well as the full length of their nose, which by the way serves a most useful purpose. If you look closely you will note that the needle nose is barbed, and this serves as a larder for the strange creature. It will stick excess food onto the barbs and keep it there until it is needed.

Spider crabs are easily collected on the reef. They are found at the edge of the coral heads usually down in a small hole in the coral. It is simply a matter of gently picking them up by the nose and placing them in the collecting bucket. But separate them if you catch more than one or you will soon have only a body with no visible means of support. They live well in the aquarium and eat chunks of fish or shrimp as well as large flakes of dry food that sink to the bottom. Large spider crabs will catch small fish in the aquarium, and with their boardinghouse reach they can grab a little fish with ease. It is most interesting to observe the spider crab in the aquarium. It uses its long legs to probe the fish or other crabs, reaching out with one leg either to ward off an inquisitive fish or to investigate it. Its eyes are equally strange, sitting at the base of its needle nose, and its legs circumvent its body like the guns on a gun turret—a remarkable creature all the way around.

There is also a hairy spider crab, not quite as large as the common spider, but equally as spectacular. It will hide in the aquarium, looking much like a wisp of algae. It is found along the West Gulf Coast of Florida in the grass. There are probably dozens, even hundreds, of different spider crabs in the oceans of the world. One species is an absolute master at camouflage and covers its body with algae, sweeping the algae back and forth when approached so as to simulate the ocean current. It is very common in Florida grass beds but extremely difficult to see because of its unique camouflage. I have caught many in a drag net when collecting sea horses. They would be simple to spot once removed from the water, because they would move about in the net. These are excellent in the aquarium but they should have a dense algae rock upon which to hide and feed.

In the deep oceans and in the colder waters, especially off Japan, there are giant spider crabs with a ten- to twenty-foot spread! What a spectacular creature for a public aquarium! In the deep parts of the sea, there are no doubt even stranger spider crabs, and someday we will see them in aquariums I hope.

Sea Spider *(Nymphon)*

Although sometimes called a spider crab, the sea spider is not a crab at all but a true spider. The common species averages about an inch in length at most, and it is virtually all legs with absolutely no body at all. The legs are like a single strand of black thread, making it a most innocuous creature. Its body is so diminutive that it is said to keep part of its organs outside fastened to one of its legs. Sea spiders are common, being found on algae-laden rocks near shore, but it takes a sharp eye to see one, because they are so inconspicuous. They live well in the aquarium if provided with an algae rock upon which to feed. They will lay eggs and reproduce, and the eggs are easily seen, because they are carried on the body, for there is positively no place that the spiders can hide them. They are also found on floating seaweed or other floating objects.

The little sea spider *(nymphon)*, greatly enlarged here, is a mere wisp of a creature with legs and no apparent body at all. It is shown here carrying eggs (probably the male).

The sea spider with eggs firmly in place. A pair were kept in the aquarium for nearly a year. It is not known whether the eggs hatched, but the parents lived on.

In the great depths of the sea, sea spiders grow to a foot or more across, which must certainly be a remarkable-looking creature. But in the aquarium, the little spiders are quite at home with their associates of minute skeleton shrimp (Caprella) and tubularia. The aquarist should have no trouble keeping them if aquarium conditions are proper.

Starfish

There are hundreds of different species of starfish and many of them will live well in the aquarium if water conditions are correct. I suspect that if anyone has difficulty in keeping the starfish, their aquarium is too clean and sterile. They are positively simple to keep. Just place them in the aquarium with a few large algae rocks and they will find plenty of food upon which to survive. I have kept dozens of them with no care whatsoever except to occasionally place a little fine

Live starfish are simple to keep in the aquarium. I have kept dozens in a single thirty-gallon tank with hundreds of other specimens. A tide pool blenny examines a starfish here. Note the brittle starfish beneath it.

Bottom view of the common starfish shows the many suction feet with which it walks. It also uses them to open clams and other mollusks.

chopped shrimp or dry food in the aquarium. I kept two dozen two- to four-inch specimens in a thirty-gallon tank for over a year with no problems. There were many snails and mollusks in the tank as well as anemone rocks, and perhaps they fed on these, but after these were consumed, the starfish still lived on and on. The only feeding would be a little dry food on the surface every few days.

If your aquarium has been medicated, your starfish may not survive, for they are very sensitive to copper or formalin. It is best not to medicate the invertebrate tank at all to be on the safe side. You can feed your starfish if you like by placing a small chunk of fresh fish or shrimp directly beneath it, but I usually don't bother feeding them at all, because they are real scavengers and can find their own food. In the invertebrate tank there is usually a lot of debris on the tank floor and the starfish will obtain some of its food from this. In addition, a little shredded shrimp or fish added to the tank will suffice as food for them. Live clams or other shellfish will also be eaten by the starfish, and a small cluster of clams or mussels from a piling near the seashore will provide it with a good supply of food.

Generally if a starfish is obtained in good condition and the tank is not kept too sterile, it will live for a long while, especially in natural sea water. A few live rocks or large algae-laden rocks will also ensure longevity for the starfish. Small starfish, one to three inches, are best for the aquarium, because the feeding problem is less. The common starfish is also the best for aquariums. The highly colored species from the reefs are likely to prove more difficult to keep because of specialized feeding, but even these would probably do better with a number of live rocks in the tank.

Basket starfish and crinoids, although extremely interesting, generally do not live too long. However, those wishing to try them should keep them in a tank by themselves in natural sea water. Give them good aeration, preferably with a small power filter, and feed heavily, especially at night, with live brine shrimp, fine chopped shrimp, and baby dry food sprinkled on the surface. Do not keep fish or other specimens with them and keep in subdued light. I feel that the successful keeping of them will be largely dependent upon proper feeding and good circulation of the water. Change a fourth to one half of the water each month and provide them with a large, algae-covered rock, propping the rock so the starfish can hide beneath it during the day.

Serpent starfish or brittle starfish are easy to keep. They will eat chopped shrimp and will even take small chunks from your fingers once they have become accustomed to feeding. Drop the chunks of food directly on their tentacles and they will coil around the food

A large burrowing starfish in the same aquarium. This specimen was over six inches long and was collected at low tide on Sanibel Island. It spent much of its time under the sand but would emerge when food was placed in the aquarium. It was eventually devoured by an angular triton; a sequence of life in the sea.

Close-up of the burrowing starfish shows numerous feet. This is one of the most active of all starfish and can move about with good agility. The sharp spikes on its arms are used in digging, for it can settle down into the sand with great speed.

The orange crinoid with its long, feathery arms is one of the most unusual of invertebrates. It comes from deep water off Florida.

A deep-water crinoid emerges between two live corals in the aquarium. Baby black angelfish nibbles on the coral.

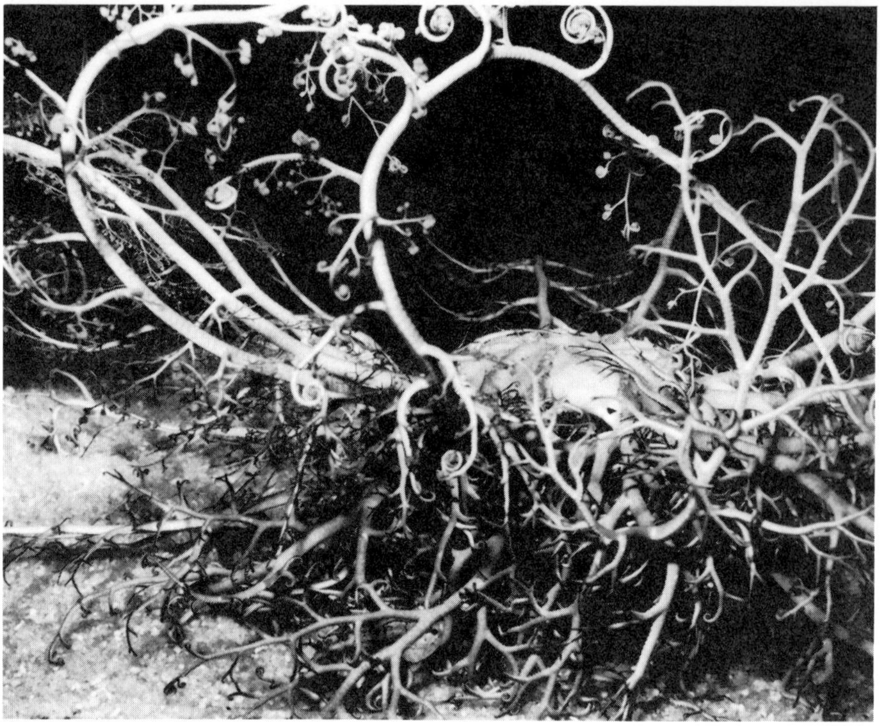

The basket starfish is an astonishing creature. Not only do the hundreds of legs and tentacles move, but the central core throbs up and down like some weird creature from outer space.

The brittle starfish is very common and is found under rocks and coral or living within dense algae as shown here. Dozens of specimens were found on this single rock, and they grew rapidly in the aquarium.

and direct the food right into their mouth. They are even more fascinating in the aquarium than an octopus, because they will come out to feed as soon as they sense the odor of food in the water. They must be handled very gently, for they are extremely brittle and will break apart in segments at the slightest touch. I have raised dozens of them by simply bringing home a chunk of rock from an area in which they live. They would appear as tiny points in the algae but would soon grow up to a size of two inches or more. If you collect any live rocks at all you are almost certain to find brittle starfish hiding somewhere inside of it, especially in tropical areas.

Live Shells

Many live shells will live in the aquarium for a year or two with little effort on the part of the aquarist. I have kept murex, tritons, cowries, and margin shells for long periods of time with a very minimum of effort. The main thing in keeping live shells is to learn which will live together, for some are extremely carnivorous and will devour another shell on sight. There are numerous shell books that will tell you about the feeding habits of most shells. *Sea Shells of North America,* by Abbott and Sandstrom, is an excellent book for this purpose.

Unusual photo shows a serpent starfish actually taking a chunk of shrimp from my hand.

A large banded tulip shell has buried itself in the sand in the aquarium. Only the tip of its proboscis is visible as the dark object at right center.

A large clam is dropped in the aquarium, and immediately the tulip shell approaches it. It could detect the odor of the clam as soon as the clam was placed in the aquarium.

The tulip shell climbs on top of the clam in an attempt to open it.

If you expect to keep a prized shell you must know which shells will attack it and keep them out of your tank, or your prized shell will simply be a meal to them. Some shells will eat shrimp, others will eat brittle starfish, and as mentioned earlier, some will even eat sea cucumbers. The large triton shell of the Pacific was found to feed on the famous crown of thorns starfish, a poisonous starfish that made the headlines of the papers and was reported to be destroying all the reefs of the world. It was later discovered that its abundance was merely a natural occurrence and that it wasn't going to destroy the reefs of the world after all. In the aquarium, you will be able to discover many new and unknown facts about seashells. They are quite amazing creatures with habits that will astound even the zoologist. The episode below will illustrate the excitement and discovery you may encounter when keeping your seashells alive instead of killing them for the shell, as many shell collectors do at the present.

Is it possible for the lowly seashell to think, reason, or communicate with another shell? Some zoologists would probably scoff at the idea, but I think it is entirely possible. Who knows what goes on in their tiny little heads?

I have kept a great many seashells alive in the saltwater aquarium for study and have been amazed more than once. Several times I have seen seashells "spin a web" just like a spider and climb down that

Two live shells in the aquarium; a large angular triton in the center has a small coral growing on its shell. To the right is a live pen shell, an interesting species from shallow water.

Meanwhile, a second tulip shell emerges from the sand.

It hurries toward the food.

web to the bottom of the aquarium. One shell in particular, a flamingo tongue, had crawled up the narrow stem of a gorgonian to a point where the gorgonian protruded above the surface of the water. I watched with amusement as the pretty shell reached the top of the gorgonian and the end of its journey. What would it do next? I expected two things. It would either somehow turn around on the narrow branch and go back the way it came or it would simply drop off the branch and fall to the bottom. But my little friend did neither. After a great deal of soul searching and considerable thought (I could tell it was thinking because it was moving its antenna rapidly back and forth), the little inch-long shell withdrew its body into the shell for what I thought was deep meditation. It would drop off the gorgonian any minute, I thought. But I was wrong. Instead of dropping off its perilous perch or turning around, the little shell began spinning a web, exactly as a spider would do. Then with equal aplomb, it lowered itself slowly to the bottom of the tank and then crawled away. Incredible! Here we have a so-called lowly creature that is perhaps more intelligent than man! (Well some men anyhow.) Seriously though, spinning a web to extricate itself from a delicate situation must have involved something more than instinct.

But that's not all. Not only can seashells spin webs, they can work together toward the accomplishment of a goal that would be difficult

At its approach, the first tulip shell moves over to the edge of the clam. There appeared to be some sort of "communication" here, for the opening of the clam proceeded in a very orderly manner.

Their antennae twitched as if sending a message, and one of the tulip shells started to flip the clam so that the other tulip could get a grip on the opposite half.

Both tulip shells are now on opposing valves of the clam. They exert a steady pressure and soon the clam opens its shell.

The tulip shell extends its head far into the clam. They take turns to feed. Note that the clam is now wide open.

The inside of the clam was completely devoured and the tulip shells moved off to again bury themselves. I have seen this remarkable cooperation of two shells combining their efforts to open a clam occur several times in the aquarium.

to accomplish by themselves. I saw it with my own eyes or I wouldn't have believed it. I even took pictures of it so you could see it too and perhaps agree with me that here we have a creature with the mental acumen of at least a chimpanzee and perhaps a porpoise.

I had placed a couple of Florida banded tulip shells (*Fasciolaria tulipa*) in one of my saltwater aquariums. I find it is much more rewarding to keep the shells alive than to kill them as many shell collectors do and have found also that most seashells are very hardy in the aquarium. You simply have to learn which shells will get along together, for some will quickly devour their shell mates. The tulip shell is a devourer and will quickly devour its subordinates, so you must not keep it with any prized shells.

But they must eat, so I obtained some large clams from a nearby grass bed and dropped one into the aquarium. The tulip shells were both under the sand with only their siphon protruding above the sand but almost instantly, as soon as the odor of the clam permeated their nostrils, both tulips were out of the sand heading straight for the clam. I wondered how they would get it open, so I sat beside the aquarium to watch. What happened was truly amazing—something one sees only once in a lifetime. I saw it three times! The seashells worked as a team to open the clam and get at their meal. One tulip shell would seat its body firmly on the side of the clam and then roll its body turning the clam over so that the other half of the clam was exposed. Then the other tulip would anchor its body to the opposite half of the clam. Next, the tulips would push against each other or pull their bodies away, forcing the clam to open. When the clam was finally weakened and its muscles tired, it finally opened enough for the tulip shell to insert its proboscis and snout within, to begin feeding. Both tulips took turns feeding on the clam until it was completely devoured. Then, after exchanging a few pleasantries, the two tulip shells wandered away and buried themselves under the sand once more to digest a well-earned meal. They had cooperated in a joint effort to obtain food.

Perhaps, if it had occurred just once, I would have noted the incident as accidental, but *three times in a row* the tulip shells attacked their food in the same manner, working together to open the clam—certainly more than coincidence. Perhaps these little, slow-moving creatures aren't so dumb after all. Just because seashells don't travel in a "school" like fish doesn't mean they don't have any education. They must know something to be able to figure out that by working together, they can open a big clam in nothing flat. This takes intelligence, communication, brains! I think we have misjudged the lowly snail. When all the cards are down, it might well be the most intelligent creature on earth.

Flamingo Tongue Shell *(Cyphoma gibbosum)*

The flamingo tongue shell is one of the most colorful of Florida and Caribbean shells and a favorite in the invertebrate family. It is a small shell about one inch in length but very beautiful. The shell itself is a rich salmon color, but the mantle is much more ornate, with rich orange and black markings, making it a prize for the aquarist or shell collector who would try to keep it alive.

It is found on the outer reefs in Florida and the Caribbean crawling about on sea fans or gorgonians. When its mantle is spread it becomes a good camouflage, for the pattern on the mantle is quite similar to the irregular holes of the sea fan. Therefore, one must look closely to find them. They are easily collected by picking them off the fans like grapes. The difficult task is locating specimens, for they are not extremely common. Should you find an area containing them, just take a few or a half dozen at most and leave the rest behind for next year and to replenish the area.

They are also found on gorgonians, for they feed on the living polyps

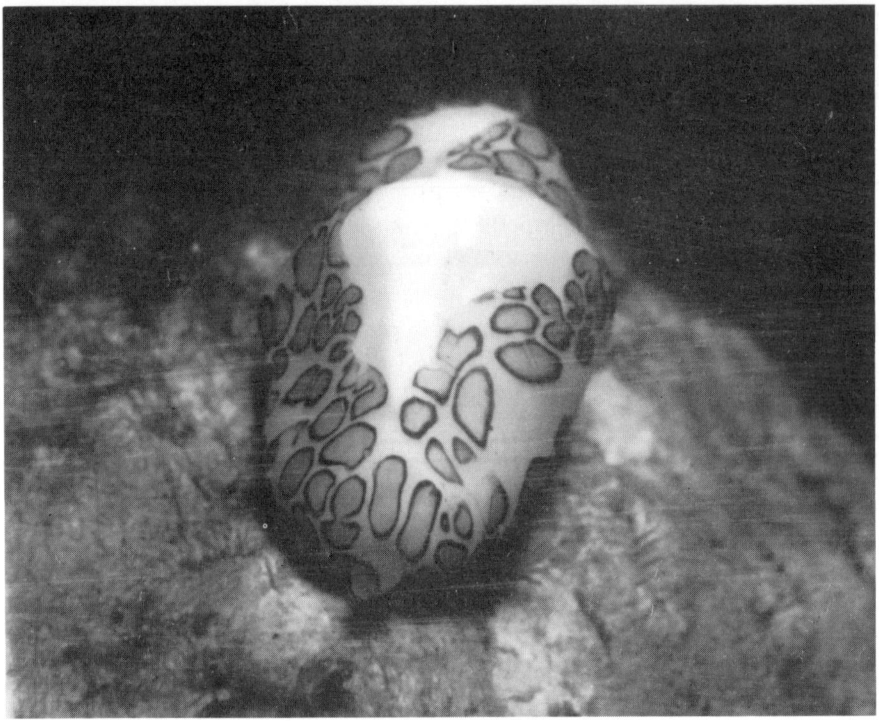

One of the most handsome of Florida shells is the flamingo tongue. It was selected as the lead photo by Dr. Paul Zahl for his classic seashell article," "The Magic Lure of Sea Shells," in the March 1969 *National Geographic*. Dr. Zahl photographed the shell in my home.

of both gorgonians and sea fans. If you collect them for the invertebrate tank, pick a half dozen small gorgonians of the same type upon which you find the shell and transport these to your home aquarium, bringing back a couple large buckets of fresh sea water from the same area as well. This will ensure food for your shells and enhance their chance for survival in the aquarium. Make certain to aerate the gorgonians well on the trip home and keep their container out of the hot sun. An algae-covered rock from the same area will also prove of benefit for the aquarium.

When you arrive home, place the gorgonians upright in the tank, pushing enough of them into the sand so they won't topple over. Use the fresh sea water in their tanks and give the tank good aeration and filtration, using a small amount of activated carbon in an outside or inside filter to keep the water sparkling clear. Place a flamingo tongue gently at the base of a gorgonian and leave it alone so it will crawl and attach to the gorgonians.

Not much is known about the keeping of flamingo tongue, but it has been kept for a year by myself and other aquarists by keeping the gorgonians with it and by changing a portion of the water each month. There are several species, easily identified by their mantle and shell coloration. Some are quite rare, like the McGinty's cyphoma. It has a different colored mantle and the shell is light purple in color. It is found in the same area as the more common shell. I have found a few of them throughout the years.

The flamingo tongue or cyphoma, as it is known in the shell trade, is a prized shell. But don't kill it if you collect it. You can keep it in a three-gallon or five-gallon tank and collect the water, gorgonians, natural shell sand, and specimens right in the same spot so that all you have to do is set up the tank when you get home. You will find much more beauty and interest in the live shell than you ever will in the dead, cleaned specimen, and later, should your live specimens eventually die, you can still keep the handsome shell.

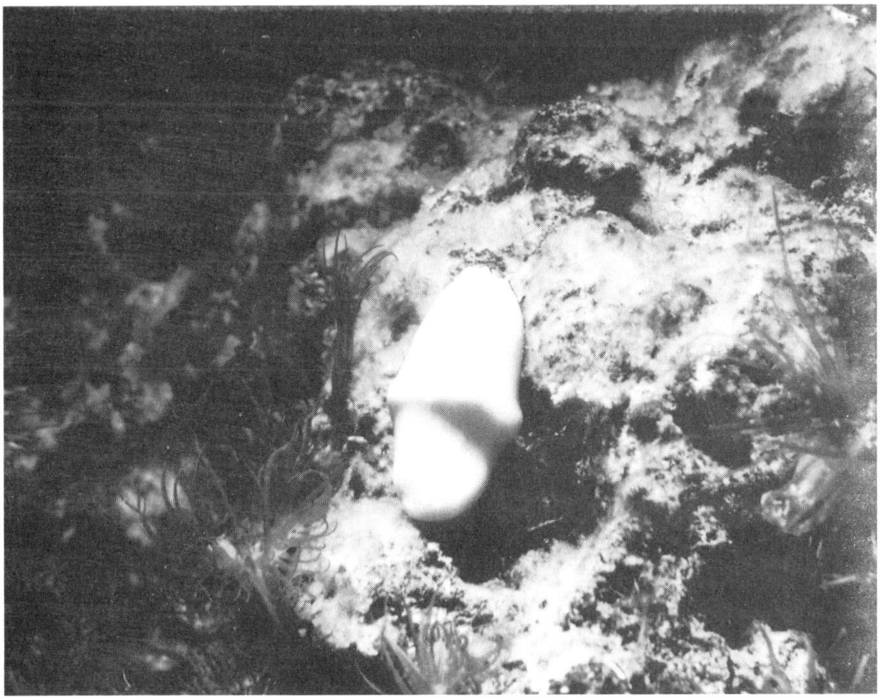

The flamingo tongue is a small shell an inch in length but quite colorful. Its mantle is extremely colorful and completely covers the shell, offering it good camouflage. But when it is touched the mantle retracts, showing the highly polished shell as shown here.

Atlantic Deer Cowrie *(Cypraea cervus)*

The Atlantic deer cowrie is a large, handsome shell with a shiny polish to its shell that looks as though it is waxed and polished. It is the large mantle of the living animal within that keeps the shell so clean, for when the creature is moving about on the ocean floor the shell is completely covered with a black mantle, giving the creature the appearance of a black wooly sponge. The collector is likely to overlook it entirely as a worthwhile specimen unless he is familiar with live seashells, for it bears little resemblance to the shiny shell in the collector's cabinet. But if you touch the black glob, it will stop and retract its mantle revealing the shiny shell beneath.

This is one of the largest of the cowries, the family to which it belongs. It grows to a length of two to four inches and occasionally even five inches with exceptional specimens. It is found in the Florida Keys and throughout the entire Caribbean. The best place to look for the Atlantic deer cowrie is under a dense clump of finger sponge in the shallow channels where the sponge is found. It will appear as a large black lump far under the sponge. It is also found in deep holes or under ledges as well as under large rocks and corals. Some collectors look for them at night using a flashlight or lantern to locate them in the shallow water. Look for this shell around bridges or rock piles extending out into the water. It will be moving along slowly and it will be simple to net it or grab it with your hands.

Cowries live well in the aquarium if fresh sea water is provided. They like several algae-covered rocks upon which to feed, but they will also eat lettuce leaves, green algae, and spinach. Because of their size, they can consume a considerable amount of food, so feed them plenty. Although they give off a certain amount of slime in the aquarium, I have never seen it cause any problems if the aquarium is properly maintained. Use a ten- or fifteen-gallon tank for them.

There is also a measled cowrie (*Cypraea zebra*) found in the same area. It too is very handsome and a prize for the aquarium. It would be well worth it to try breeding and raising these beautiful shells in the aquarium, and it should not be an impossible task. Use a large tank of thirty to fifty gallons and fill it with several choice algae rocks, first picking off all crabs and shrimp and other shells. Then place several of the adult cowries in it and feed them well. Give the tank good light, changing a little of the water each month. If all goes well, the cowries will spawn in the aquarium and then you will have the challenge of rearing them. Many seashells have spawned in aquariums in the classroom as well as in the home. (See "Measled Cowie," *Salt Water Aquarium Magazine* Sept.-Oct. 1970.) This is a marvelous field for research and investigation, where nearly everyone has the first chance to breed and raise live seashells in the aquarium. The cowries are considered by many to be the most beautiful, and there are cowries throughout the world that are stunning in color. Some are extremely valuable, and are worth thousands of dollars.

The Scallop

The most colorful of the scallops is surely the flame scallop (*Lima scabra*) with its brilliant, fire-red interior. It is a pure delight in the aquarium and of great interest. It will spread forth its fiery red tentacles and open its valves to their fullest as it siphons food and water. But its shell is very drab and unattractive. A shell collector who has never seen a live specimen will miss the real beauty of the flame scallop.

Flame scallops are found in the Florida Keys under flat rocks and coral on the ocean side of the Keys. The outside of the shell is sharp and rough and some collectors call it the file shell because of the rough surface. It is best to use gloves when removing them from a rock. There are several color phases of the mantles ranging from fire

The most active and colorful of Florida scallops is certainly the flame scallop, *(Lima scabra)*, shown here with its fiery interior. I first gave it the common name, which has stayed with it through the years.

The orange bay scallop is equally fascinating. Its many eyes can be seen peeking through the grooves in the shell.

red to orange, tan, or white. They grow to a length of about four inches and live well in the aquarium. Feed them large quantities of newly hatched brine shrimp, fine dry food sprinkled on the surface, and occasional feedings of fine, macerated shrimp or fish. Change a little of the water each month and do not keep crabs, drill shells, or fish that may attack them, in the same aquarium.

The bay scallops are nearly as fascinating to keep in the aquarium. These have bright blue eyes, easily seen with a sharp eye or a hand-held magnifying glass. Feed them the same as the above and they will last a long while. There are numerous species of scallops on both coasts that will be interesting in the aquarium. The thorny oyster (*Spondylus americanus*) would be the prize. It is found in deep water in the tropics and a diver with scuba gear is needed to collect them. It is found attached to old wrecks. But there are many much more common oysters and scallops that can be obtained by anyone who journeys to a rocky area of the seashore. Some grow to good size, being nearly the size of a dinner plate. It would certainly be interesting to keep one alive in the aquarium. There are thousands upon thousands of different seashells throughout the world and most of them will live well in the aquarium, including even the giant clam.

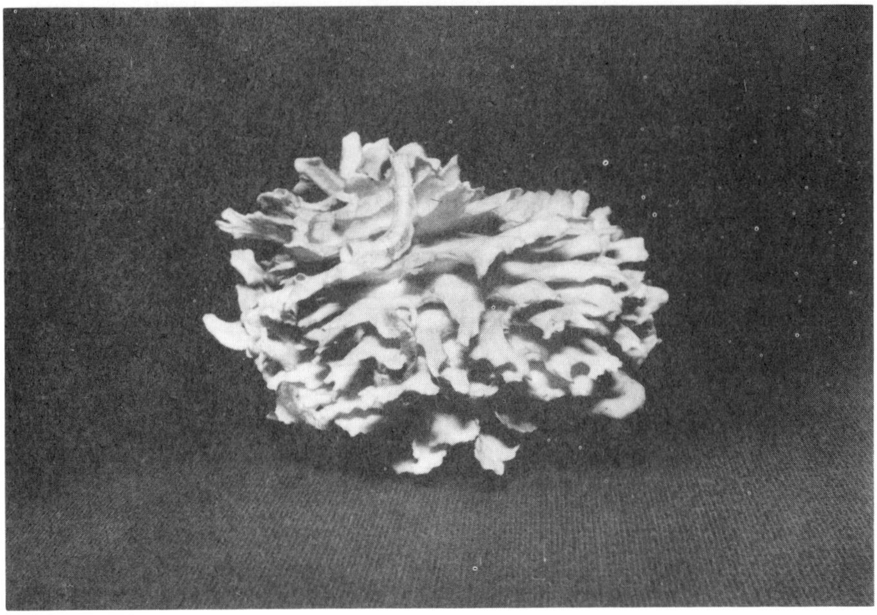

The jewel box (chama) is a prized shell to the collector because of its delicate colors and leafy appendages but it is even more of a prize in the aquarium, for it lives extremely well. It is often brought in on a piece of live rock or coral, where it is tightly attached.

The giant clam (tridacna) also has an interesting shell, but the live animal is far more spectacular with its brilliantly colored mantle. Small specimens could certainly be kept alive in the aquarium.

The biology student visiting the seashore on a collecting trip should arm himself with a large, ten-gallon plastic bucket instead of a jar of formalin and make an attempt to bring the seashells back alive. There is much more to be learned from a live specimen. Some of the jewel boxes, like the leafy jewel box (*Chama macerophylla*) of Florida, are among the most intricate and beautiful of all shells. They are found firmly attached to rocks, sometimes many of them clustered together. The rock and all are collected and transported to the aquarium where they will live on indefinitely. In the aquarium they can be studied alive and their rate of growth may be determined. On a worldwide basis, the variety of seashells would stagger the imagination. If only the shell collector would keep them alive instead of killing them! The astute conservationist who finds a rare shell could even keep it alive in his aquarium for a few months, then return it unharmed to the sea, knowing that someday he might find it again and perhaps recognize it by a special marking on its shell!

The Mantis Shrimp

Mantis shrimp are very common in semitropical or tropical seas. The name is derived from the long front claws, which resemble those

Razor sharp claws of the mantis shrimp or squilla have earned it the name *thumb splitter* by shrimpers, who collect many in their nets.

of a praying mantis. Although the mantis shrimp is seldom seen by the underwater observer, this is only because it is shy and retiring, for it is actually very plentiful. One scientist from the University of Miami Marine Laboratory reported that net hauls from the Gulf of Panama were completely filled with mantis shrimp and that the bottom of the sea must be covered with them.

In Florida and the Caribbean, they are found under rocks, finger coral, or inside sponges. Pick up a large rock with numerous holes and it is almost certain to contain at least one mantis shrimp. They average one to three inches in length for some species. Others may be ten to twelve inches, and some are reported to be over two feet long.

They are easily maintained in the aquarium. Give them a few algae rocks upon which to crawl and feed and they will make most interesting aquarium residents. They will look at you from behind the rock with their strange eyes. They will perform acrobatics, doing a somersault, swim upside down or backward. They may decide to rearrange

Nearly all snails, periwinkles, and shells will live well. Some teed on each other. Others will eat chunks of fish or shrimp as well as vegetable matter. The author has placed as many as fifty assorted shells in a single aquarium, where it becomes "survival of the fittest."

the tank for you and will dig down into the sand like a terrier after a rat, or they may carry sand in their huge claws and move it about the aquarium, dropping it like an earth-moving machine.

But don't get your fingers too close to them, for they can lash out and inflict a severe cut, which has earned them the name of "thumb splitters" by commercial shrimpers who catch many of them in their nets. They will eat chunks of fresh shrimp or fish, which should be dropped near them or hand fed to them on the end of a slender rod if you wish to tame them. A large specimen is intriguing, for it has many unusual habits and such an unorthodox appearance. Yet it is seldom displayed in public aquariums. It has, however, become a favorite to the aquarist who has discovered it. Don't keep it with coral shrimp or prized fish and other specimens, for it may eat them.

Scarlet Cleaner Shrimp *(Hippolysmata grabhami)*

It is difficult to say whether this spectacular shrimp is more colorful than the banded coral shrimp. Perhaps it should be placed in a class of its own. It is certainly a beauty with it flame-red body, huge white antenna, and broad white streak down its back. I have collected a

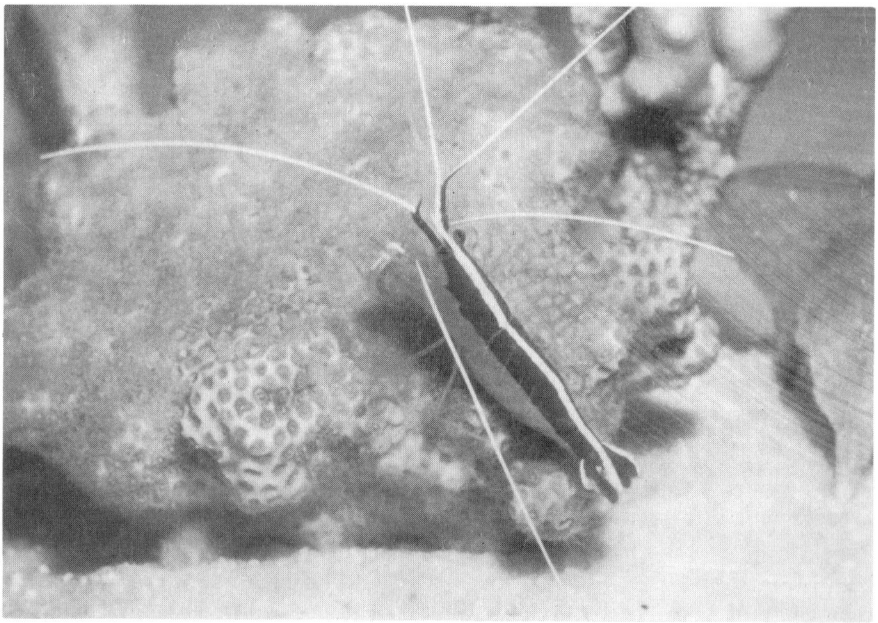

No other shrimp is quite as colorful as the scarlet cleaner shrimp with its red-and-white-striped back and huge antennae.

number of them through the years but I have never found it numerous. In fact, I would say it is quite rare, at least at this writing. Once in a great while, you will find one or two under a ledge in similar habitat to the coral shrimp. You will first spot their long white antennae, for they will turn to face you. They are easily netted by driving them down into the net with a rod or gloved hand. If there is a mated pair, collect both of them and keep them together in the aquarium to see if you can breed them.

Because they are rare at present and expensive if you purchase them, it would be best to keep them in an aquarium to themselves. A five- or ten-gallon all-glass aquarium will be fine. Use natural sea water if possible and include several algae-covered rocks. Make a ledge under some of the rocks so they can hide beneath it. Feed chunks of fresh shrimp, fish, and dry food and change a little water each month. They will live a long while under proper conditions but don't keep them with predators, for as soon as they shed their skin you will lose them. They vary in size from three inches to seven inches including the antennae. Some green algae would be desirable.

The Bumblebee Shrimp

Those who like a small aquarium will take special delight in this strange little shrimp, for a full-grown specimen is about the size of thumbnail or smaller. This is a most interesting creature, fairly common if you know where to look for them. Their small size results in most people overlooking them entirely. They are found under algae-covered rocks in the shallow reef areas of the Florida Keys and in the Bahamas. They have a strange, erratic swimming movement of short quick dashes, and they will dart quickly back into the rock from which they have been removed, so you must look quickly if you are to find them. Sometimes they can be found under sea urchins or under basket starfish, where they may live in symbiosis.

A small, two-gallon glass tank with an algae rock is fine for them. Prop it up so they can get beneath it but so that you can still see them. Feed them small chunks of shrimp or fish. They are tiny but interesting. I have caught them as small as a pea yet they have lived a long while in the aquarium. Keep predators away from them.

Other Shrimp

There are thousands of different types of shrimps throughout the world. The common bait shrimp is available in many places and in the aquarium it has many interesting qualities. Give it several inches

Barely discernible, the diminutive bumblebee shrimp ranges from pea size to fingernail size.

of sand under which to hide and you will be able to observe it in near-natural conditions. At night it will emerge from the sand to scavenge for food. Shrimp are true scavengers and will eat most anything including other shrimp. Feed them dry food, chopped shrimp, fish or scallop, and algae.

Red cleaner shrimp are found in great numbers in rocky areas along the Florida East coast. These will live together in groups and should be kept in a tank with several algae rocks tilted upward so they can hang on its under surface. Deep-water shrimp are found off the coast of Florida in several hundred feet of water. They are bright red with huge antennae and should make interesting aquarium specimens if you can obtain them from a commercial fisherman or research vessel working in the area. Rose-petal shrimp (*Hymenoceros elegans*) from tropical seas are sometimes available to the aquarist and are among the most decorous and ornate of all sea creatures. Should you be fortunate enough to obtain a pair, keep them in a tank to themselves so they won't be attacked by other specimens. They are said to feed on small starfish.

The oceans of the world contain an abundance of shrimp and crabs that can bring education and delight to the biology student as well as to the home aquarist. But at this writing, the Department of the

Interior is trying to pass legislation that would prohibit the importation or interstate shipment of all invertebrates under a catch-all title called "injurious wildlife." If this passes it will be disastrous to the student of marine biology, who would be denied many interesting and unusual specimens. I have written letters to the president, vice president, and the Interior Department protesting this unneeded and unwarranted legislation and have encouraged others to do the same. But should these laws be passed, they should be repealed, and if every reader would take time to write his congressman about it, perhaps something can be done to remedy the situation. In my letters, I stated that everything in the world is injurious to something else but that doesn't mean it should be banned. If our government *really* wants to protect us why don't they force the auto makers to make a *safe* car, a nonfracturing gas tank, protect us from drugs and crime, make a fireproof home, build an electric car or a hydrogen-fueled car that would make its own hydrogen on demand from a tank of water, prohibit huge gasoline tankers from moving on the same highways with fragile autos. I can think of a million things that would be more "injurious" to us Americans than a live seashell or a coral shrimp, yet our govern-

Anemone shrimp live within the stinging tentacles of various sea anemones. Several different species are shown here, including the orange and white "clown shrimp" on the lower right of the anemone.

ment is trying to ban this very thing. WRITE THEM ABOUT IT! Perhaps it is the Department of the Interior itself that is "injurious" to a free America!

Anemone Shrimp

In addition to the above shrimp, there are also many little shrimp that live within the tentacles of the sea anemones. These vary in shape and color. Some are almost transparent, and I call them ghost shrimp, for you can see right through them. Others are more substantial with powerful little claws that allow them to cling to and walk right across the face of the sea anemone completely unaffected by its powerful stinging cells. One species in particular I call the clown shrimp (I may have been the first to discover it many years ago). It is orange and white, the same color as the percula (*A. ocellaris*) clownfish of the Pacific, and it usually lives on a large stinging anemone, found in the Florida Keys. Not only is its color similar to the Pacific clownfish, but it moves its body in the same weaving motion as the clownfish. It will close up its tail and swirl the tail around like a snake charmer would move his flute. Perhaps it is trying to lure a fish to its doom when it does this.

Those who wish to keep and study the anemone shrimp should collect both the anemone and the shrimp at the same time or order them from the collector. It is best to transport the shrimp separately from the anemones unless one is very careful in handling them. The anemones and shrimp should be kept in an aquarium to themselves to avoid predation, and the tank should be fed good quantities of newly hatched brine shrimp as well as shredded shrimp and scallop. The anemone itself may need a chunk of fresh fish fed directly into its tentacles. Anemone shrimp should make a fascinating study for the marine biologist or zoology student, who certainly could uncover many new facts about them by careful observation in the aquarium.

Pacific Clown Anemone *(Stoichactis sp.)*

The Pacific clown anemone is easily the most popular of all anemones, for to watch a small group of clownfish snuggle down into its tentacles is truly a wonder of the aquarium world. Ten or twenty years ago, this was a near impossibility for the average aquarist, because the clown anemone was extremely rare in this country and was available only to the largest public aquariums at a very high price. A hundred to several hundred dollars was often paid for them. The price has gradually come down from fifty dollars to as low as ten dollars at

Close-up view of the Pacific clown anemone reveals velvet texture, but it contains powerful stinging cells that will kill nearly all fish, except for the clownfish that live within the tentacles.

the retail level. It will probably stablize at around ten to twenty dollars unless it is prohibited from entering the country.

It is a sensational creature. It is very active, constantly changing size and shape or moving from one part of the aquarium to the other to get into better light or into an area of suitable current. I do not know the longevity record for this interesting anemone. Sometimes it does very well, other times it slowly wastes away. Since it is such an active anemone, I suspect that the reason it does not do well is because it is underfed. In the ocean it receives a constant supply of plankton and small fish, so this must be duplicated in the aquarium. Give it good light, fresh sea water if possible, a firm rock upon which to anchor, and immense feedings of newly hatched brine shrimp, fine chopped shrimp, and chunks of fish dropped directly into its interior. Never medicate its aquarium and give it a good circulation of water, either with a power filter or by opening the air bubbler stems of the sub-sand filter so they make a strong movement of water in the aquarium.

This is a stinging anemone, so handle with care. Don't let it touch your tender upper arms or it may give a severe sting. Usually it won't

The Pacific clown anemone *(Stoichactis sp.)* is the most prized of all anemones. This specimen spread out to over a foot tall in the aquarium.

A large Pacific clown anemone *(Stoichactis sp.)*, nearly a foot across, in one of my tanks.

hurt the tough skin of the palms unless your skin is quite tender, but it definitely will burn if it touches a tender area. The clownfish should be kept with it but first sterilize them in a separate container with copper to be sure they don't bring disease to the tank. If the clownfish show signs of disease after they are in the aquarium, you can treat them with copper by simply scooping up a gallon of water from the tank in a plastic measuring pitcher and fastening this to the inside of the tank with two clothespins. Insert an air stone into the container, then add copper to it. Leave the clowns in this for a few days, then dip them out and return them to the aquarium, using a small net. But be sure no copper gets into the aquarium itself. If the clowns still are not cured, simply return them to the copper solution. When the fish have been cured, remove the copper solution and discard it, replacing the lost water with a gallon of fresh sea water. This sometimes works better than a separate hospital tank, for the fish are treated right in their own water, which is kept at the same temperature. Also it elimi-

Like a bright flower, the Florida pink anemone decorates the aquarium. This is probably the most popular of anemones because it is usually available. A blue angelfish swims below.

The Florida pink anemone has graceful tentacles and is found in many colors from bright pink to red or multicolored.

nates the need for a hospital tank in many instances.

It may be necessary to remove your anemone and place it out in the sun, but a good aquarium light will usually furnish sufficient light. A little water is replaced each month with fresh sea water for best long-range results.

The Florida Pink Anemone *(Condylactis passiflora)*

The Florida pink anemone is the most common anemone available to aquarists or the collector who vacations in Florida. It is found in very shallow water near shore and is easily collected if it is picked from the grass beds. It may be gently picked up by hand by reaching under the base to remove it. Or it can be collected from a boat with a dip net by dipping a corner of the net under it and scooping it up. It should never be collected in rocky areas, for it is nearly impossible to remove it from a rock without damaging the base. In a good area, it is possible to collect a hundred in a few hours when one has become proficient in collecting them, but this is for the professional collector.

The average aquarist or student should only take a few, for that is all he is likely to need.

Look for them down in the Keys of Florida on the ocean side, in shallow channels where the current flows across grass beds. There will usually be a little sponge or finger coral in these areas, which will be your clue that you are in the proper area. The best time to find them is at high tide, for they are out full and easy to locate. At low tide, they are liable to be pulled down into the grass. Sometimes you can find them on the grass flats in small pools at dead low tide. But don't waste your time collecting them on the rock shelfs, for it is so much easier to get them in the grass.

They live extremely well in the aquarium, especially if it contains fresh sea water. Give them good light and copious feedings of brine shrimp, and fine dry food as well as a little fine chopped shrimp spread directly on the tentacles. Avoid large chunks of fish or shrimp. They may need an occasional sunbath outside in the bright sun if they don't do well for you. Usually they will live for years with little extra care.

The wondrous anemone rock with hundreds of anemones is perhaps the most interesting thing you can keep in the invertebrate tank. Yet it can be collected in less than a minute. It will live for years in the aquarium with little care.

Proper water conditions, good current, and light are all important to their survival.

Anemone Rock

Anemone rock is probably one of the most interesting formations that can be kept in the aquarium. It is simply a piece of dead coral rock that is covered with tiny, dwarf sea anemones. The anemones are light brown to black in color and are mature at a half inch. A rock six inches across may contain hundreds of them. They are extremely hardy and will rapidly multiply in the aquarium, dividing in two when they reproduce. The aquarist can start out with a single rock hosting the anemones and easily fill an aquarium by placing clean rocks in the tank. As the anemones reproduce, they will cover the new rock. I have raised thousands of them this way and ship the live rocks to public aquariums and universities as well as to individual hobbyists.

The anemone rocks seem to purify or stabilize the water in a salt-water aquarium as I have reported numerous times in *Salt Water Aquarium Magazine* and in other books. An aquarium may be plagued with disease or microparasites and harmful bacteria, making it unsuitable for coral fishes. But add a few anemone rocks and in a few weeks the tanks will undergo a change and become completely safe. I suspect that either the anemones filter out harmful organisms from the water or they add an essential substance to the water, an enzyme or vitamin or a combination of both. I don't believe this has anything to do with the so-called bacterial cycle of the tank, for in tanks in which they were used, some of the tanks had been set up well past the year mark and held numerous fishes. In this length of time, there was surely sufficient bacteria in the sand to establish the bacterial cycle, but it would not be until the introduction of the live anemone rocks that the tank would really flourish.

Anemone rocks are easily collected right from shore in the Florida Keys. Look for a grassy area where there is not too much current and where there are small rocks near the shore. The anemone rocks may be covered with dense algaes and other growths and the anemones will hardly be apparent when you lift the rock out of the water. They will retract and will appear as tiny black bumps, slimy to the touch. You can check to see if they are anemones by placing the rock back in the water and observing them to see if they put out their tentacles. If the water is too murky, place them face up in a bucket and observe them after they have settled for five or ten minutes. (Anemone rocks are not coral and should not come under Florida's new coral law.)

The brown dwarf anemones are found in the Florida Keys and west

Orange coral from Puerto Rico is one of the most colorful of corals and lives for years with proper care. I obtained these specimens from John Dinga, a professional collector.

The dwarf anemone matures at a half inch and multiplies rapidly in the aquarium. A large rock can contain dozens or hundreds of them. Anemone rocks seem to purify the water in the aquarium, often ridding it of disease and parasites.

coast of Florida. On the east coast, perhaps all the way up to the Carolinas, there are green dwarf anemones of a different species. These are found in the coquina rock, where they adhere to the underside. They appear as brown lumps when the tide has receded. Many of the rocks are almost completely covered with them but usually on the underside. They are easily collected in tide pools at low tide. Look for a rocky area, then look beneath the rocks for slimy half-inch bumps. Place a small sample in clear water to check for anemones. The rocks may be picked up if they are small in size, but if they are attached to a large area, then you must pry a small section loose with a small pinch bar or pipe. Select a piece about ten or twelve inches long, since this is easily transported home in a bucket. The green anemone rock can be found high and dry at low tide but it will usually be under a ledge where it can remain damp. It should be transported in water to protect the many different forms of life in it.

Anemone rocks are easily kept alive. Place them in clean sea water with sub-sand filtration and a good circulation of water as described for other specimens and feed heavily with live brine shrimp. Light feedings of fine chopped shrimp or clam are also recommended. An anemone rock will give you hours of endless fascination, and there will be small crabs, sea spiders, and perhaps nudibranchs living in the same rock. These are among the hardiest of all anemones, and so long as you keep the salinity down to 1.025 or less, they will live on for years and years.

Live Gorgonians

Some species of gorgonians or alcyonarians live well in the aquarium. The alcyonarians are often called *soft corals* by the marine biologist and include many interesting forms, including the dead man's fingers and the sea pen. The gorgonians are referred to as horny corals because of their interior horny construction. But many are similar in appearance and it would take a specialist to identify one from the other in many instances. The scientific names of the individual species may be known on a worldwide basis, but in my opinion, they are extremely confused, being "borrowed" from books from one country to the next and translated to different languages. I would not hazard a guess as to the exact scientific name of any species, because I would not wish to contribute to the confusion. It would be a simple matter for me to place a scientific name under a photo, but the name in all probability would be incorrect, for the polyps of many species look very much the same. The serious student who needs the correct scientific name of specific gorgonians, alcyonarians, and corals should look to the special-

An alcyonarian with huge, starlike polyps fully extended is an amazing sight in the aquarium.

ist in the field who is well informed on the correct nomenclature. General books, scientific magazines, and even scientific institutions supplying the specimens may not be one hundred percent correct in their identification.

Gorgonians that live best are generally the thick-stemmed type with large polyps. When retracted, the stem is three-eighths to one-half inch thick. Specimens less than a foot high are most easily transported alive. Handle them gently when collecting them, for if you break the skin on the stem, they will not live well. Chip a small section of the sea floor when collecting them and place just one or two upright, in a bucket of clean, aerated sea water. Keep them out of the hot sun. Place them upright in the aquarium and give good light, aeration, and filtration. Feed live brine shrimp and fine macerated shrimp or clam. Change a little water each month. They should live easily past a year or longer with proper care.

The purple gorgonians, said to be the alcyonarian, dead man's fingers, is one of the hardiest of the species. When handled properly, it will blossom forth into a magnificent display of fluffy polyps that will prove of tremendous interest. I have kept them for years and placed many on exhibit. But they must be collected without damage

A live gorgonian with its polyps fully extended in the aquarium.

The same specimen with polyps retracted after being touched.

The purple alcyonarian, a type of gorgonian found in shallow water, lives extremely well in the aquarium, putting forth its polyps in a shower of beauty. A Pacific angelfish *(Pomacanthus semicirculatus)* nibbles on algae growth brought in with the alcyonarian.

Extreme closeup of an alcyonarian reveals the strange polyps with newly formed polyps growing between the large polyps. A ten-power magnifyer will reveal the same view.

to their inner body or they will die and foul the entire aquarium. They will give off a red-colored solution that will kill everything in the tank and create an odor of 7.8 on the nasal offense scale (pretty bad).

The larger gorgonians will also live well in the aquarium if they are transported properly. I have kept three-foot specimens alive by transporting a single specimen in a fifty-gallon plastic drum. They make a spectacular display. The main thing in either collecting or shipping them is to place just one to each bucket or shipping box with clean sea water. This way they have a chance of making the trip alive, but if you put several in a bucket or bag they will soon suffocate. The slender gorgonians will also live in the aquarium but are not as hardy. They give off a heavy slime when disturbed. Those who wish to try them should select small specimens. Sea fans are also difficult to keep, as mentioned earlier, but I would encourage the student or aquarist to try keeping them, for this is the only way we will learn about them. Pick a small specimen attached to a piece of rock and place the rock and all in a bucket of clean sea water. Keep the water out of the sun and aerate all the way home. The fan will live for a while in the aquarium and perhaps with proper feeding of brine shrimp and fine macerated shrimp or clam it will flourish and grow. It would be quite

A purple alcyonarian in "bloom" with its polyps extended. A young jack-knife fish swims in the center.

A small sea fan in the aquarium. This specimen was attached to a small coral head, so both coral and fan were collected together, keeping the firm base of the sea fan intact. Beautiful gorgonians are also in view.

an accomplishment to grow one in the aquarium up to full size. You would be able to study its rate and manner of growth.

The Sea Pansy

The sea pansies of pennatulaceans are well known to the marine biology student, but few people have kept them alive in the aquarium when compared to the masses that have been pickled for science study. It is a shame that more of them are not collected and shipped alive, for certainly much more can be learned from the live specimen. They are quite hardy in the aquarium and change form constantly to the delight and surprise of the aquarist. When the polyps are fully extended, they may be three or four times the size of the retracted specimen, and one wonders at first if it is the same specimen.

Sea pansies live very well in the aquarium. Place them in fresh sea water or artificial that is well aged. Give them good aeration, filtration, and light. Feed newly hatched brine shrimp and fine macerated clam, shrimp, or fish. Change a little water each month. They should live for months or years. But do not keep crabs, shrimp, starfish, or fish in with them, because these may attack them and kill them.

The sea pansy with most of its polyps retracted. When washed up on shore it would appear even less conspicuous, for the stem would be shrunken and curled up around the body. It would look much like a purple toenail.

The same specimen with polyps extended for feeding. The body has also expanded and the stem has been anchored into the sand.

Sea pansies are easily collected with a small trawl a short distance from shore on the east coast of Florida, northward. They can be collected in huge quantities once a bed of them is found. They may also be collected at low tide after a period of stormy weather. They will be found right at the tide line, but it will take very sharp eyes to detect them. They will appear as a flat, mouselike object the size of a thumbnail with a small curled tail and a rough, sandpaperlike covering on the body. If you collect them as soon as the tide has fallen, they will still be alive. They should be immediately placed in fresh sea water to revive them. Once the sun has reached them, they will dry out and die, so only specimens close to the tide line should be collected. I collected the specimens portrayed in this book at low tide at New Smyrna Beach and they lived fine in the aquarium, requiring no special care.

Sea pens are found in deeper water but sometimes can be found at low tide. They are of great interest because they give off a bright light when they are touched. Some of them grow to lengths of six and eight feet, which would be quite a specimen for the aquarium. They retract down into the mud at low tide but can be removed by scooping the

The best time to look for live specimens if collecting from shore is during a period of stormy weather. Rough seas will wash sea pansies, starfish, live shells, algaes, and nudibranchs ashore. This beautiful view was taken at Venice, Florida. Numerous fossil shark teeth may be found here at low tide.

mud beneath them to avoid damaging the animal itself. They are found off both coasts, so anyone can collect them once he learns what he is looking for, since a glob of slime is often difficult to identify by the beginner.

The light given off by the strange creatures will vary in color and intensity with the species. It is said that in the deep sea, huge colonies of sea pens give off so much light that they look like subterranean cities when viewed from a deep-diving submarine—another wonder of the sea.

Provide the sea pen with several inches of sand and a little mud. Give it good aeration and feed heavily with brine shrimp, macerated clam, and fine dry food. It will prove an interesting addition to any aquarium, and since it is found in northern waters, the cold-water aquarist will find it of equal fascination.

The Sea Squirt

Sea squirts are called tunicates or ascidians and are found in all

A pair of sea pansies, one with polyps extended and the other with them retracted. These appear to be mating but it may have just been an accidental meeting.

seas, from the tropics on up into the colder waters of New York and Maine. They are interesting in the aquarium because of their immense diversity in shapes and colors as well as body forms. Some look like a piece of an old shoe, while others are brightly colored, almost as ornate as the coral fishes.

They are easily identified as sea squirts because of the openings or siphons at the top or sides. These are tiny spouts that take in and expel water as the creature feeds. When a sea squirt is first removed from the water it will often squirt a small stream of water from one of these openings, which has given the creature its name. This will be your clue to collecting them. Pick up a clump of spongelike marine growth and if water squirts from it, you will be able to discern the sea squirts if they are present. Some may be black or brown and appear like a wrinkled potato when they have expelled their water. They may be solitary or clustered together. Some species live in large colonies.

Keeping them requires no special care. Simply place them in the aquarium and feed live brine shrimp and other foods, the same as you would corals. They will siphon out the fine particles of food from the

Extreme close-up of the sea pansy shows polyps similar to the gorgonians. The live specimens can be studied in great detail in the aquarium, for they live well.

water and require little care. I have kept them in two-gallon aquariums along with a complement of fish, and they lived on indefinitely. They seem to condition the water in an aquarium, making it healthier for fish, perhaps by their fine siphoning and filtering action.

Nudibranchs

Nudibranchs are among the most colorful and exquisite of all the creatures in the sea. They are the butterflies of the ocean and range in all seas. The colors and patterns they present would fill a book in itself, and someday an astute aquarist-photographer will come out with a full book on just the nudibranchs of the world. It would be an interesting book, but would have to be in full color. Meanwhile you can find a great many of them portrayed in a number of books and magazines. They are classified as seashells, for they are simply snails without shells, but the resemblance to the common snail is indeed misleading, for most nudibranchs are creatures of delicate beauty.

They are found in all seas from the very shore to the great depths as well as on the surface of the oceans. Possibly the Great Barrier Reef has the most colorful species, but this is difficult to ascertain, for I have found exquisite specimens right in the Florida Keys. Also some very colorful types are found in the cold waters of the Pacific North-

A spectacular orange and black nudibranch I collected near Miami, Florida.

A strange flatworm crawling across tubularia in the aquarium. They are found in same area as the nudibranchs.

west. There are thousands of species and many yet to be discovered. They vary so much in shape and color that rarely are two of the same type even photographed, except for the *Glaucus atlanticus,* one of the most interesting of all. This dark blue beauty is found swimming on the surface of the open sea. But it is often found close to shore or even washed up on the beach when the wind and tide are pushed from a certain direction.

Glaucus atlanticus from the Atlantic or Glaucus marina from the Pacific are considered pelagic nudibranchs, but when the wind is blowing from the right direction, many of them will be washed ashore. The purple sea snail (*Janthina janthina*) , another interesting specimen for the aquarium, is often washed ashore at the same time as is the deadly man-of-war jellyfish (*Physalia*) . These poisonous jellyfish, easily recognized by their clear, light blue float and the long strings of stinging cells, should be avoided at all cost, for they can be deadly. There are a numb r of deaths caused by them, and their sting is extremely painful. Do not handle them or go in the water when they are sighted. I was almost killed by one when a full sized specimen plastered itself to my back and throat. It is a deadly creature, make no mistake about it!

The purple sail, velella, and porpita umbella are also blown ashore

Nudibranchs are snails without shells, and many are of exquisite beauty. This handsome specimen was brought in on a piece of algae-covered rock.

at the same time, and glaucus has been seen feeding upon the latter, which may be its principal food. A few years back, a magazine from Africa ran a picture of the tiny "sea dragons," as they called them in the caption. They had no idea what they were and asked for identification, which I promptly supplied. Many people see the little "sea dragon" but have no idea that it is a seashell or nudibranch.

Nudibranchs are easily collected on the coral reefs or rock patches but you must look for them specifically or you won't see them. Their delicate color blends in closely with their surroundings. They come in all shapes and sizes. Some are tufted with poisonous quills on their backs. These can sting, for they carry nematocysts from stinging hydroids that they have consumed. Others have tufted plumes on their backs or frills of delicate finery. To find them you must cover every inch of area on the reef and look for a little "flower" among the algae growth. Sometimes if you move your hand rapidly near the algae, fanning it, the nudibranchs will be washed loose by the water current. They are very common but they are not easy to find. I have caught

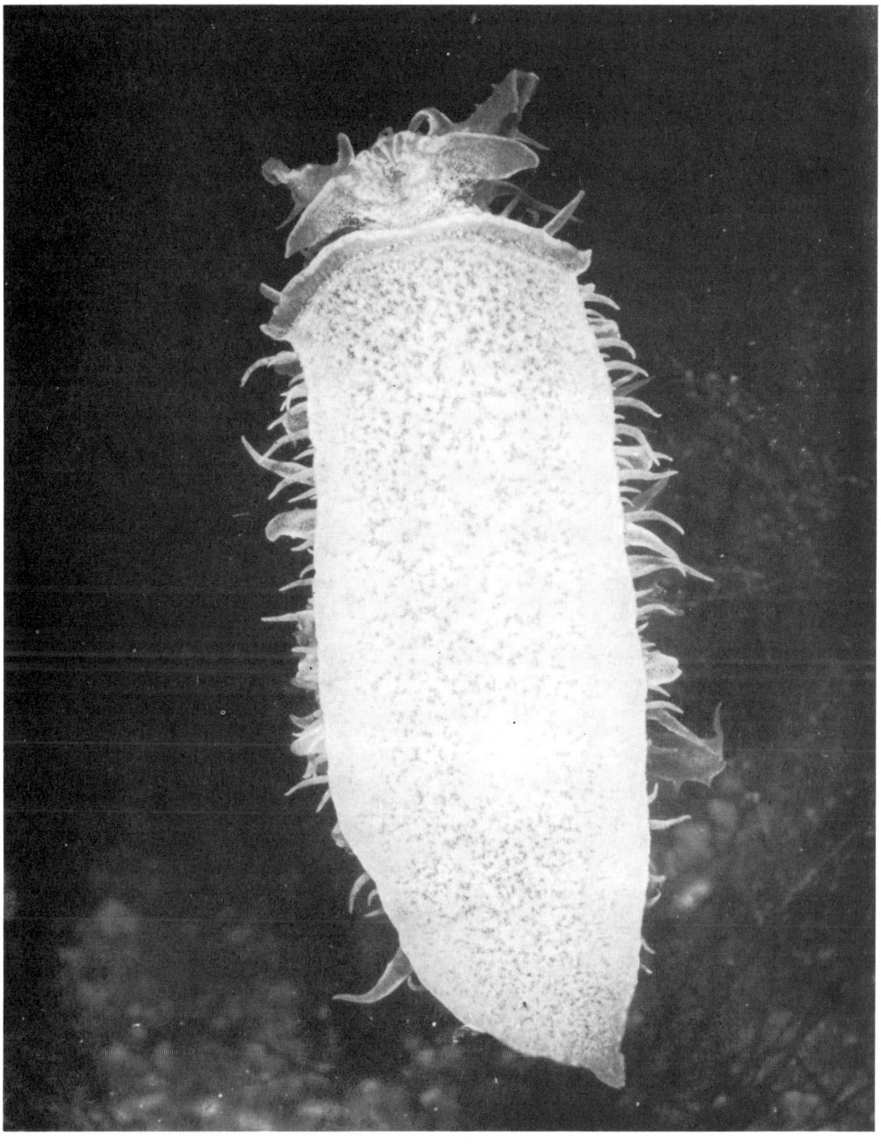

This large nudibranch or sea slug was three inches long and was picked up from shore at Sanibel Island. It lived in my aquarium where it fed on algae.

thousands of them through the years, all colors of the rainbow, and never cease to be amazed at their color and beauty.

They range in size from less than a half inch to at least two or three feet long. The littles ones are very insignificant on the reefs but the big ones two feet long are easily seen, even from a distance. I have found large ones that were jet black and would swim through the water like a giant bat. When handled, they give off a red ink that will stain the hands and is said to be mildly poisonous. The large nudibranchs graze upon the algae beds in shallow water and "herds" of them can be seen eating the algae like so many cows. They eat a tremendous amount of the algae, gulping it down almost as fast as a rabbit. Should you decide to keep some of the larger types, make certain to collect a good supply of the algae at the same time so they will have enough to eat.

The smaller nudibranchs will live well depending upon the species. The specialized feeding of many are not known, but generally an algae-covered rock combined with an anemone rock will furnish at least some food for them, combined with a little dry food and fine chopped shrimp. Give good light to encourage algae and replace the algae rocks as needed. Minute sponge growth and hydroids are usually found on the algae rocks, and certain nudibranchs will eat this as well as other growths. I have had nudibranchs spawn and have raised the young in the aquarium by providing them with rocks and an occasional change of water.

Nudibranchs are so beautiful and fascinating that some day aquarists will specialize in keeping them. Aquarium shows will be held to display and compete, not for the finest guppies, but for the best nudibranch.

The Sun Jellyfish *(Cassiopeia)*

Jellyfish are always interesting to observe but unfortunately few do well in the aquarium. This doesn't mean that you should not try to keep them, for you should at least try. Many will live for at least a few days and you can derive much interest from them in that time. Perhaps it is because they are pelagic that they do not do well. A doughnut-shaped aquarium of clear plastic may help to keep them longer, because it would direct them around in a circle so they would never stop. You can have such a tank manufactured by a plastic company or you can make one of sorts by simply fastening a sheet of thin plastic into a cylinder, then fashioning a smaller cylinder in the center and place this in the aquarium. A power filter can be used to move the current in one direction so that the pelagic jellyfish will move

The sun jellyfish *(cassiopeia)* is one of the few jellyfish that will live in the aquarium for any length of time. Dense algae is caulerpa.

around and around, never coming to a stop. In addition, the water must be well filtered and aerated. Huge quantities of live brine shrimp as well as fine shredded fish, shrimp, clam and fine dry food may help to keep the creature alive.

I have kept many varieties alive including the man-of-war jellyfish for weeks but never really made a determined effort to keep them longer. But there is one jellyfish that can easily be kept in the aquarium. This is the *cassiopeia or sun jellyfish*, an interesting species from Florida and the Caribbean.

It is found in very shallow water, often under the shade of the mangrove trees, where it is found in large numbers. It ranges in size from a half inch across to about a foot across. Small specimens two to four inches are best, because they are more easily transported. They are simply collected by scooping them up carefully in a net, but handle carefully because they are easily damaged. Also, they have a slight sting that may affect some people more than others.

Keep in a well-aerated tank and give good light. Feed live brine shrimp, fine dry food, and macerated clams. A slight change of water will also prove beneficial. They are easy to keep and will live several months or longer with proper care.

The man-of-war *(physalia)* with its gas-filled float is common at Florida beaches at certain times of the year, but don't touch: its sting can kill, especially when the tentacles are concentrated as shown here. It can kill even good-sized fish. I was almost killed by one when a large specimen plastered itself to my back and throat.

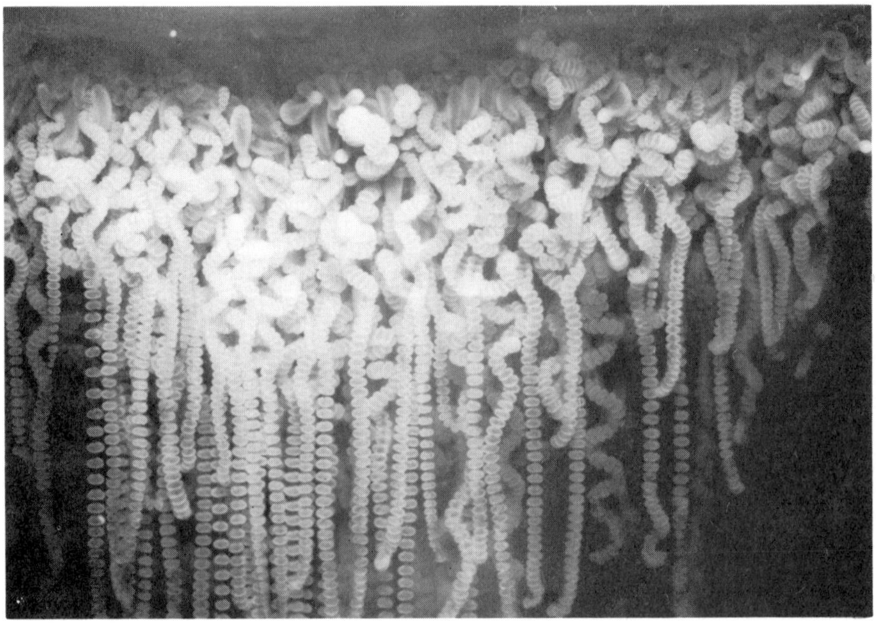

Closeup of stinging cells shows thousands of cells each containing a poison more lethal than that of a cobra. The man-of-war can be kept in the aquarium only a short while. No sand should be used and the specimen should be whirled about in a circle.

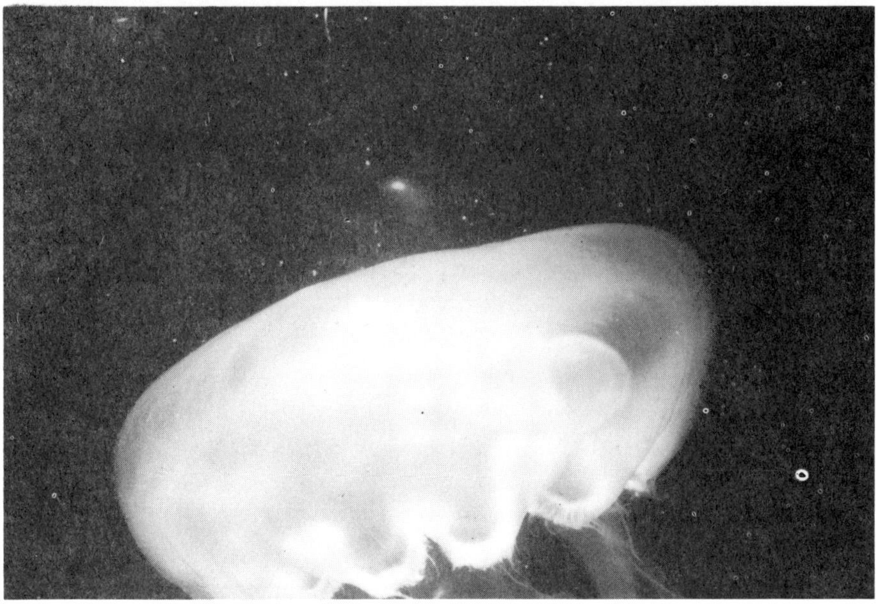

Most jellyfish are difficult to keep in the aquarium for any length of time, but many will live for several days to a few weeks, like the interesting specimen above.

The man-of-war jellyfish (physalia) living in the aquarium. Although it doesn't live more than a few days or weeks in the aquarium, a great deal can be learned about it in that time. The aquarium should be rounded so that the jellyfish will not come to the end of it. No sand should be used, and a power filter may be used to keep the water moving in a circular motion. Sunlight will help sustain it.

The Sea Urchin

There are many species of sea urchins in both northern and tropical waters so that the aquarist can easily collect them for his aquarium. Unfortunately in California and perhaps Oregon, it is illegal to collect them from tide pools. On one hand we have the "urchin kills," where skin divers kill thousands of urchins to protect the kelp beds. On the other hand we have laws prohibiting the taking of the urchins (both in the same state) . It doesn't make much sense. But for those who can collect them, the urchin is an interesting specimen for the aquarium. They come in many colors and shapes.

Keeping them is simple. Just place several in a ten- or twenty-gallon tank with fresh sea water and a few large algae rocks. They are scavengers and will eat nearly anything that falls to the aquarium floor. Shredded shrimp, clam, fish, and dry food will be eaten. The algae rock will furnish extra food to balance their diet and you should have no trouble keeping them in properly maintained aquarium, particularly if you select small specimens.

The long-spined urchin (*Diadema*) is easily the most spectacular of the sea urchins. It is a real porcupine! It can sting, too. The long

The long-spined sea urchin (diadema) is easily the most spectacular and makes a fabulous aquarium specimen. It should be handled with a flattened net to avoid damage to the specimen as well as to your fingers.

The common sea urchin will live well in the aquarium if it is not kept in a sterile tank. Here it lives with anemones and corals.

spines are needle sharp and will easily pass clear through your finger or hand if you strike one hard enough. I once had one pass right through my finger and protrude three inches out the opposite side. Consequently they should not be handled except with a net. Even then, to avoid breaking the beautiful spines, they should not be scooped up in the net. Instead, hold the net closed and scoop them up on the flattened net like you would a pancake on a pancake turner, moving rapidly, for they crawl quite fast. If you try to pick one up by hand, they will turn the spines toward you and may impale your fingers. However, they can be picked up by sliding your hand beneath them and holding them in the palm of the hand, but this takes practice and nerve. I have picked them up this way but prefer a net, for those needles do hurt, especially when they stick under your fingernail and come out at your elbow! (A little exaggerated, but they do hurt, believe me.)

The long-spined urchin is a large species. A full-grown specimen may be a foot in diameter. It is fascinating to keep and lives quite well with proper care. Algae rocks are needed and chopped shrimp or fish may be placed directly under the urchin so it can find it easier. Give good aeration and filtration, changing a little water each month.

A common green sea urchin in the aquarium will live for months or years. The green caulerpa to the left had to be removed to another tank, because it proved to be a favorite food for the urchin, who devoured it faster than I could collect it.

The club-spined sea urchin is also very hardy. It will live for years if provided with an algae rock upon which to feed.

A colony of tubularia is of special interest to the biology student but how many of them have observed it alive? It is easily kept in the aquarium, where it may be studied with a low-power microscope or hand-held lens.

This small ten-power lens is ideal for the aquarium. Its clear sides admit light for viewing and it is adjustable so that it can be focused just inside the aquarium glass.

Small to medium-sized specimens are usually more practical, especially for the inland aquarium, but at any size, they are interesting.

The club-spined urchin is also worthwhile in the aquarium. The Florida species averages about three inches across at most, but some of the Pacific club urchins are much larger, approaching half a foot or more. They are said to be poisonous, so they must be handled with extreme care, but I have been stung by them often in the Philippines without serious effects. However, the Florida species is not poisonous. I have handled thousands of them with no problems. They live well in the aquarium. The only requirement is an algae rock and occasionally a feeding of chopped shrimp.

The Small Creatures

The biology student will find great interest in *live* tubularia, astrangia, caprella, the tiny skeleton shrimp, and nymphon, the sea spider. All of these interesting creatures can be found near shore on algae-covered rocks, especially in the cooler climates, and during the season when they are available. Keeping and studying the *live* specimens are much more interesting and are not difficult at all. It simply requires the acquisition of a few large rocks containing the desired specimens and an aquarium with fresh sea water.

The tubularia is an interesting pale pink hydroid that is found in dense clusters at certain times of the year. It will live for a considerable length of time in the aquarium, so that it can be studied with a magnifying glass by the student or interested aquarist. It is often found in accompaniment with the tiny skeleton shrimp (*Caprella*), a weird little creature about 1mm long to as long as a half inch, which is sometimes collected in vast numbers by simply picking up a rock from the area in which they live. Obelia is also found in the same area and the stinging hydroid hydractina may be found on hermit crabs within walking distance of shore.

It takes a specialist to identify many of these minute organisms and my friend Warren Prince, a biological collector, has acquainted me with many of them since they are so small that I would have passed them by. But in the aquarium they come to life in many unusual forms and the magnifying glass exposes many "monsters" that would hardly be seen with the naked eye. Thus you have a "living" laboratory, certain to unfold new discoveries about the very tiny but important creatures that live in the sea. How they grow and reproduce can easily be studied in the aquarium, but it would be an impossibility to watch most of these near-microscopic creatures under the sea.

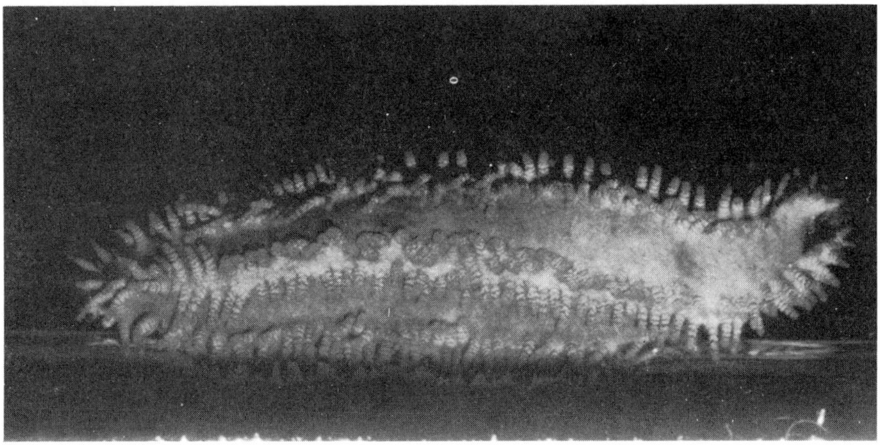

Even sea cucumbers are worthwhile in the aquarium, and there are many different forms. This dwarf species rarely exceeds two inches and lives extremely well. It is found off Florida's Gulf coast.

The aquarium keeps them alive nearly as long as you would need them and longer in most instances, and with practice we will eventually learn how to keep all sea creatures alive and in good health for indefinite periods.

I have merely touched upon some of the interesting invertebrates. There are many more and some that have not even been discovered, especially those from the deep sea. There are strange corals from the great depths that would be of great interest to any aquarist, myself included, and especially to the scientist or serious student of the sea. There are abysmal sea cucumbers, sea lilies once thought extinct, and deep-sea octopus that look like something from another world, which they are—the undersea world.

Keep any of them in the aquarium and you should have learned at least something, even when keeping the ordinary shallow-water species, for we still have a lot to learn about them.

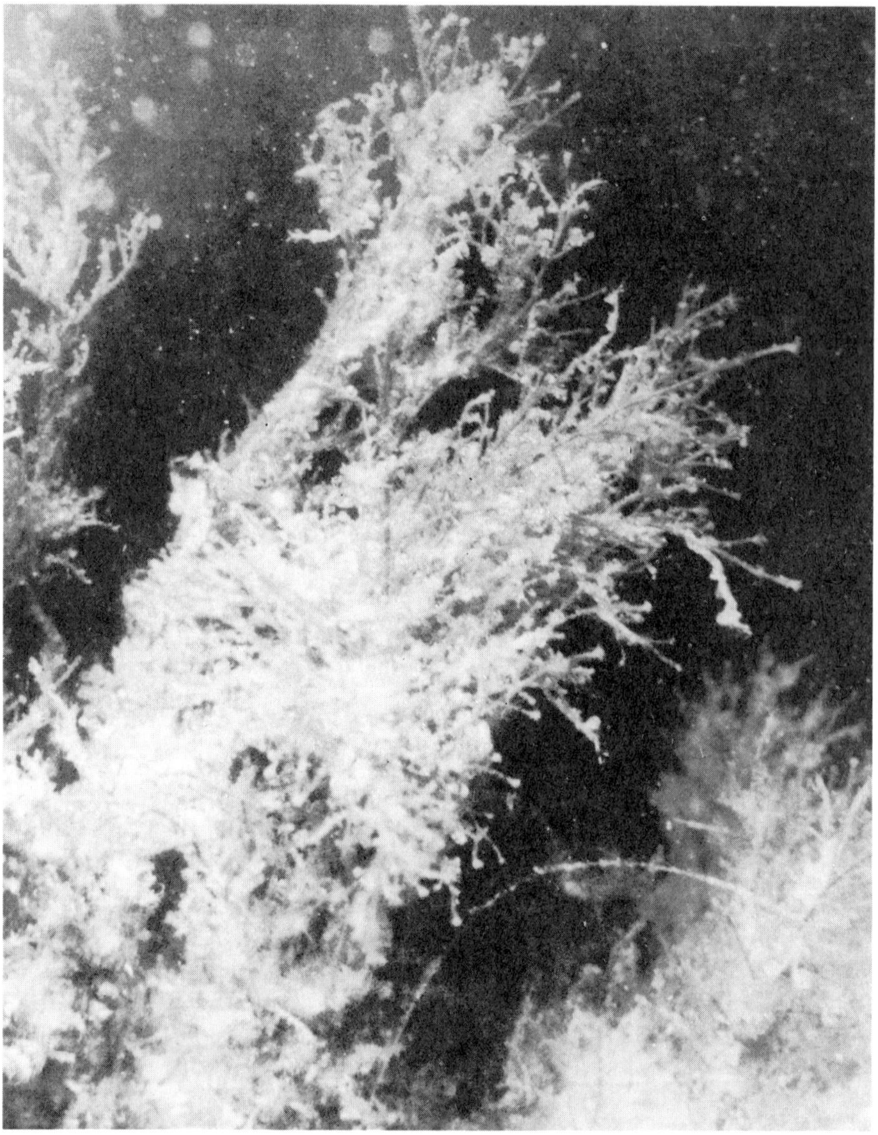

This moss-covered rock contains thousands of caprella, the skeleton shrimp, which are barely visible to the naked eye. Several can be seen here along the outer edges, but a magnifying glass should be used to see them clearly.

Closeup of same photo shows a skeleton shrimp at right, center.

8
The Mangrove Swamp

THE WONDER AND MYSTERY OF A MANGROVE JUNGLE CAN NOW BE TRANSplanted to your living room. This fascinating area of the tropical seashore is little understood by man as an essential ecological balance in the chain of life to fish, invertebrates, crustaceans, and shore birds. Numerous important forms of life make their start among the protective roots of the mangrove tree, in the shallowest of waters close to shore. Water birds nest and breed in the mangrove.

Yet man, largely ignorant of this, continues to remove and destroy the valuable mangrove from his shores. He calls it progress. It is really devastation! With this book, I will try to reverse this trend of mangrove destruction so that the public will become aware of the importance of the mangrove swamps and estuarine areas. It is essential for the very survival of our fisheries, the perpetuation of game fishing, the survival of millions of shore birds, and the procreation of the lobster and crab industries. This is more than enough reason for total protection of our mangroves.

Did you know that the mangrove swamp, so eagerly bulldozed and destroyed by oceanfront "developers," is actually a nursery for billions of marine organisms essential to the chain of life of all fish, birds, and crustaceans found along our shores? Baby lobsters by the millions thrive and mature in the protective jungle of roots and shade provided by the mangrove tree. Crabs of all types make their home in the mangrove jungle during various stages of their growth. Numerous fishes, the snapper, the grunt, grouper, parrot fish, even the mighty barracuda, live in the mangrove roots in their juvenile stages. Hundreds of different species live in the mangrove at various times in their life cycle. Millions of birds seek safe refuge in these interesting trees. In fact whole rookeries of pelicans, spoonbills, and herons can be found

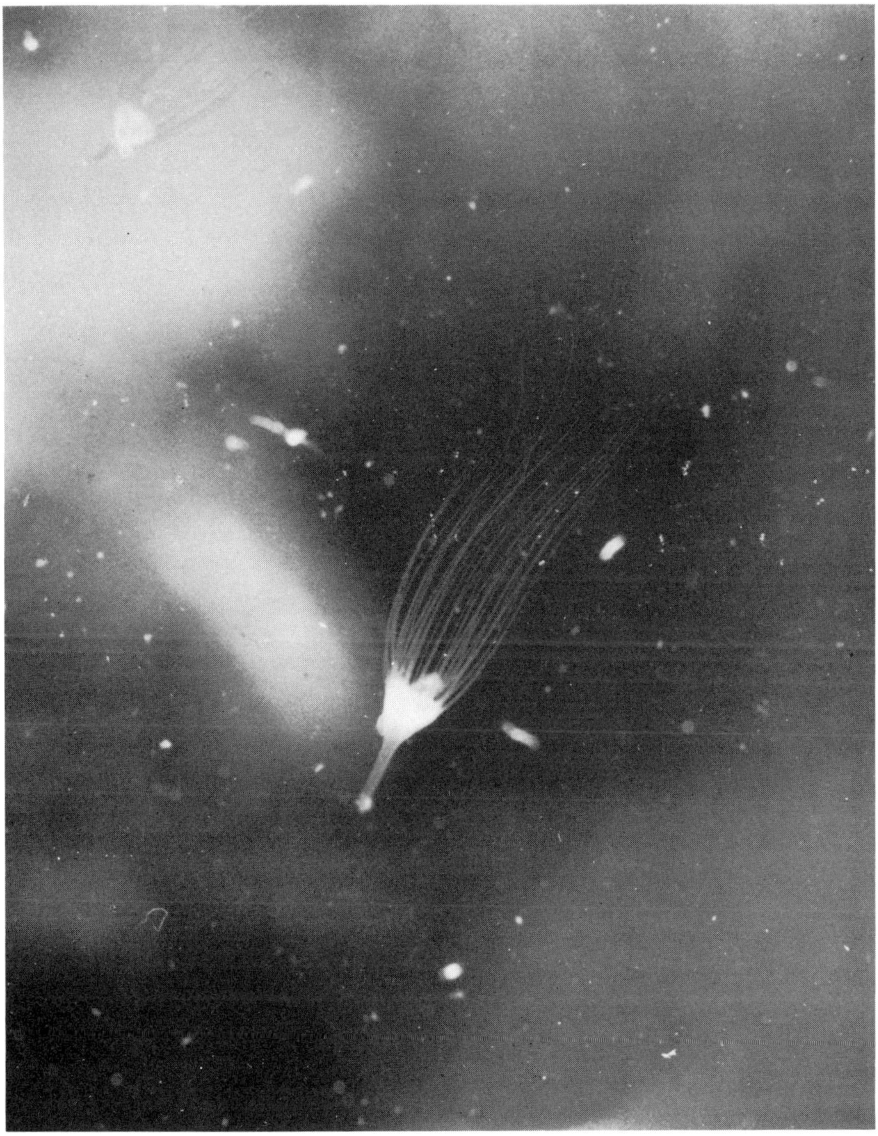

Like a comet streaking through space, these interesting organisms materialized in the aquarium and attached themselves to the glass. Tube worms were also born in the aquarium.

The colonial anemone is found in very shallow water near shore. When placed in the aquarium, it will bloom like a hundred bright sunflowers.

The coral polyp astrangia is also of interest to the biology student, who often works only with the dead specimen. It is very hardy and lives well in the aquarium.

The valuable mangrove is a nursery for billions of marine organisms and essential to the chain of life for fish, birds, and crustaceans along our shores. Yet man continues to destroy it.

nesting among the mangrove. Isn't this enough reason to preserve the mangroves along our shores? Indeed it is!

There should be a law to protect the mangrove from further destruction. Let those who would build and develop our vanishing ocean front leave eighty percent of the mangrove intact. They are not only destroying the mangrove in front of their own property, they are destroying the fisheries of the entire area, which is unfair to the public. The oceanfront should belong to all of us. The best law of all would be to prohibit *all* future building within five hundred feet of the shore so that the area could remain forever free and undeveloped. In many areas, one can no longer visit the sea shore, because it is all hotels or private property. Oceanfront property owners should be paid the current price for their land then return the land to the public for their enjoyment. This would protect our mangrove areas as well as our vanishing oceanfront and preserve them forever for the public to enjoy.

Meanwhile, until this is done (if ever), let the oceanfront developers build *around* the mangroves. They may find that this would actually benefit them, rather than hinder them. First of all, the mangrove is a native tree, already growing in the area. It will do better than any tree that could be planted there. It won't cost a penny and it is a most interesting and decorative tree. Mangroves will add beauty and interest to the waterfront. They will provide shade, a home for the birds and fish, and will even protect a waterfront home from wind, rolling waves, and debris during a storm. All this is certainly sufficient reason to leave them along the waterfront. Perhaps before man has totally destroyed the mangrove, he will realize their value and begin *planting* them along his waterfront instead of removing them. He will then find improved fishing around his dock, natural and interesting shade trees, a soft, springy bumper to protect his boat from the rough shore line should it slip its mooring. These are just a few advantages of keeping the mangrove trees intact along our oceanfront. There are many more.

How many creatures live in the mangrove? The answer is millions, even trillions if you want to include the smaller organisms. The list is almost endless but here are a few:

Mangrove snapper
Mutton snapper
Lane snapper
Grunts
Parrotfish
Snook
Tarpon
Channel bass
Weakfish
Jewfish
Grouper
Crocodile
Rabbits
Spiny lobsters
Crabs
Oysters
Coral shrimp
Snails
Shrimp
Barnacles
Sponges
Pelicans
Spoonbills
Herons
Raccoon
Wildcat
Barracuda

This is just a sampling of the wide variety of life that lives or gets its start in life in the mangrove jungle. If the mangroves are destroyed, it effects the lives of all the above and many more. One purpose of this book is to bring about an interest and awareness of the mangrove tree so that it will not be wantonly destroyed. The seeds and seedlings that will be collected to grow mangroves in the home will not in any way harm the mangroves along our shores, for they are cast-off seeds that otherwise would be either raked up and hauled away as trash in

A mangrove bean or seed pod with roots at the bottom and budding stem at the top. They are washed ashore by the millions in the Florida Keys and throughout the Caribbean. Plant them upright in the sand and mud and watch them grow.

developed areas or would slowly rot in the mud along the shore. Mangrove trees are extremely prolific and put out millions of seeds. In nature, only a few can survive, otherwise the entire world would be engulfed by them. But they do need a specific, partly submerged area in which to grow. When this area is leveled and filled by man, the mangrove is lost forever to the area.

Planting the Jungle

This book will tell you how to grow your own mangrove jungle right from seed. Mangrove trees grow very slowly, so if you are in a hurry, please pass up this project. Your mangrove seeds won't sprout into luxurious growth in a week or two. In fact it may take several weeks for your plants to sprout their first leaves! But if you are over-anxious, you could start your jungle with sprouted seeds or even half-grown plants so that you will have something to look at while you are waiting for the full-grown mangrove jungle to mature.

You will need the following items for a mangrove jungle.

1. An aquarium, ten to twenty gallons capacity, preferably the new, all-glass type or a plastic aquarium. If you intend to confine your jungle entirely within the tank, use an aquarium twenty inches to thirty inches tall; then you can put a light over it.
2. A light for the aquarium so you can view your display in the evening. Also it must have light to grow.
3. Aquarium sand, about two pounds per gallon, twenty gallon tank will take about forty pounds, plus ten pounds of mangrove mud.
4. Coral rocks: five or six, small fist-sized rocks. Three or four larger rocks, six to eight inches across.
5. Sea water or sea salts to make sea water.
6. Mangrove seeds or plants, ten to twenty pounds mangrove dirt.
7. Plant fertilizer (Nutrisol is excellent), peat moss.

Setting up and maintaining the mangrove swamp or jungle is a very simple task. You won't need any special skill or tools. The above equipment is just about all you will need, except for occasional replacements or additions to your jungle. Just follow these simple instructions and you should have success with your swamp or jungle.

1. Set the aquarium on a firm base. Cement blocks are excellent. They are strong and level. You can build them up to any desired height. Place a sheet of plywood on top of the blocks, then set your aquarium on the plywood. If using the new, all-glass aquariums (which should be used if possible because they won't rust or corrode), place a sheet of Styrofoam, one inch thick, on top of the plywood. Then set aquarium on top of it. This will lessen the chance of the glass breaking.

A mangrove jungle with coral rock and a number of mangrove trees. It must have good light to get started properly.

2. Locate the aquarium so that it will get a couple of hours of sunlight a day, if possible. Do not put it where it will get too much light or green algae will form in the water area. If you live in an apartment with no sunlight, you can still have a jungle, for it will grow by artificial light. Select a special plant-growing light.
3. Place some of the small coral rocks in your aquarium, making a small semicircle with four or five rocks in the front of the aquarium. This will be your oceanfront reservoir.
4. Wash the aquarium sand thoroughly in a bucket. Mix in one cup full of peat moss and ten pounds of mangrove dirt with it after it is well washed. Then scoop it up and place it in the back portion of the aquarium to a depth of about four or five inches. Do not put any in your little pond in the front of the tank. Keep mangrove dirt or mud well to the back away from the pond.
5. Place several large coral rocks on top of the sand several inches apart, leaving open sand areas between them.
6. Now insert the mangrove seeds down into the sand, pushing them close to the bottom of the tank. The fat or thick end of the bean is the root area and this should be pointed down. If the beans already have roots, take care not to break them. Push them gently to the bottom and cover them with sand so they remain upright.

Use about a dozen mangrove beans or small plants for an average twenty-gallon aquarium, about two dozen for a fifty-gallon tank. They can be planted quite close together because they grow very slowly and all of them may not sprout. You will need about thirty pounds of sand for a twenty-gallon tank, or perhaps even forty pounds. Figure about two pounds of sand per gallon capacity of aquarium.

7. Now add saltwater to the aquarium. Pour in enough so that the little pond in the front of the aquarium is about an inch or two deep. Sprinkle a small amount of water around the plants so that it will soak down to their roots. The depth of water in the pond should not be more than two or three inches deep depending on the size of the aquarium. There must be two inches of sand protruding above the water because the plants cannot grow totally submerged. Also, do not cover the bean too much. Only about a third of the bean should be planted in the sand for best results. If you do not have sea water, you can make artificial sea water from a synthetic sea salt. I use Reef Salts in all of my work. Just mix it with fresh water until it is clear. Then let it sit about a week in a clean glass jug before use. One pound of salts will make about three gallons of sea water. You can purchase the salts at most aquarium stores.

Mangroves will grow in any good soil. I grew these trees near my house in Miami, far from the ocean.

That's all there is to it. Your mangrove jungle is now completed. All you must do now is sit back and forget it except for occasional slight maintenance. It won't grow fast, but who knows, five years from now, you may have a good-sized jungle. Just imagine, a genuine mangrove jungle growing right in your house!

Maintenance of the Jungle

After the jungle has been set up for a few days, check the water level in the little pond at the front. It may be necessary to add more water, because some of it may be absorbed into the sand and plants. Add sea water so that the pond is a couple of inches deep. Be certain that it doesn't flood over the sand area in back. About half of the sand should be above water at all times. If sand is four inches deep, two inches should be above the water level of the pond in front. Observe the following simple steps.

1. Turn on the aquarium light or strong floor lamp directed down into jungle so that it gets six to eight hours of light each day.
2. Once a month, siphon out, or dip up with a clean glass, most of the water from the pond and place it in a clean container overnight or until the following afternoon. This will serve two purposes. It will clean the bottom layer of water in the pond, especially if you siphon from the bottom, for it will take out most of the dirt and allow drainage out from under the deeper sand area. This will reduce accumulation of gases that will build up at the bottom of the tank. Also it will allow the lower root area of the plants to breath a little air. This is a natural occurrence. In the ocean, the tide does this every day as it falls and rises.

   If you have fish or other marine life in your little pond, you will have to leave an inch or so of water in the tank so they can swim, but if the tank has no swimming inhabitants, take out most of the water while you are siphoning it out.

   Allow the water to settle quietly in the container, and then the following day, you can dip it up with a clean glass and return it to the pond, leaving the dirt at the very bottom of the container. If the water is very dirty and in poor condition, simply discard it and replace it with clean, fresh sea water or artificial sea water. It is best to discard the water at least three or four times a year and replace it with clean water. You are dealing with a small quantity of water here, usually just a few gallons, so it should not present any great problem. Keep a five-gallon jug of fresh sea water or artificial sea water near the aquarium or use three or four gallon jugs for the same purpose. The latter will be easier to handle. When you use them up, simply refill them with fresh solution.

Making Artificial Sea Water

If you are located inland and cannot obtain fresh sea water, you can easily make your own sea water. Purchase a good quality of synthetic sea-water salt. Reef Salt is the brand I use, but most popular brands are satisfactory. For general aquarium use, one pound of salt will make three gallons of water, and if you anticipate keeping little saltwater fish, sea horses, or other "critters" in your pond, you should mix it to this proportion. But if you are keeping nothing in the pond except for a few crabs, you can use a weaker mix of one pound per five gallons of water. In this case, you simply put a pound of salt in a five-gallon jug and fill the jug with fresh water. Then shake the jug back and forth until the salts are completely dissolved. If using gallon jugs, simply divide the pound of salt into five equal parts and place one portion in each jug, fill with water, and shake well until dissolved. That's all there is to it. Let the sea water set for about a week before using it so that it will age a little. If you want to be more scientific you may check the water with a hydrometer made especially for this purpose. It should measure 1.015 if you are not keeping live fish in the pond or 1.025 if you do intend to keep fish in it. Mangrove tress are not fussy and can tolerate a wide range of salinity or salt content. They are extremely hardy. They sometimes grow in extremely salty areas and sometimes grow in brackish water, which is water that is partly or nearly all fresh water.

How Soon Will They Grow?

Your mangrove beans or seeds should start to sprout leaves two to four weeks after you plant them. Do not despair if they don't sprout right away. Mangrove seeds are extremely hardy and sometimes they take a long while to sprout. As long as the bean is firm and green, it should eventually sprout. However, if it turns completely brown or black and becomes soft and wrinkled, it is most likely dead and should be discarded from the aquarium. Never start out with beans in this condition. Also, make certain that the bean is planted properly. The thick end should be planted down into the sand, the fine, pointed tip is where the first leaves will sprout at the top.

If the sand becomes very dry at the top level, once a week you may dip up a small amount of water from the pond section and pour it gently around the plants to dampen the sand. Once a month, sprinkle the plants with *fresh* water, very lighly, to rinse dust from the leaves and to moisten the plants. This simulates rain that would occasionally fall out in the wild.

Fertilizing the Jungle

Once a month, sprinkle a teaspoon or two of fine peat moss around the base of the plants. You may also add a teaspoonful of regular lawn fertilizer, pellet type, to help stimulate the growth of the plants. Do not overfertilize. Use a liquid fertilizer. Nutrisol will do fine if available (most hardware stores have it). The plants grow very slowly and need little fertilizer until they are very large. They will obtain considerable nutrients from the sea water. Also, the peat moss will tend to keep the aquarium acid, which is a natural condition of the mangrove jungles. Little is known of the full plant requirements of the mangrove in captivity, and experimentation with special plant

The common fiddler crab makes a fine inhabitant for the mangrove swamp. They are easily collected at the seashore.

foods and fertilizers may produce faster growth and more luxurious trees. Your local plant club or nursery will be able to assist you here.

Miniature Inhabitants for the Jungle

There are many small, interesting creatures that may be added to your mangrove jungle to give it more life and to simulate an ecological balance. It will add a little movement and life to an otherwise totally quiet scene and bring more interest to the whole project while you wait for it to grow. All specimens should be obtained in the very smallest size, for the whole project is a miniature form of a mangrove jungle, which out in the wild would occupy thousands of square miles. Even a single tree would be twenty to thirty feet tall and have an overall spread of ten to twenty feet or more. Below are listed some of the more common and easily obtained specimens for your jungle.

Fiddler crabs	Tree snails
Mangrove crabs	Gammarus
Hermit crabs	Pistol shrimp
Baby lobsters	Porcellan crabs
Pigmy sea horses	Blue crabs
Snails	Horseshoe crabs
Oysters	Land crabs
Barnacles	Land hermit crabs
Pipefish	Starfish
Gambusia (mosquito fish)	Sea squirt
Mollies, saltwater	Sea urchin
Cassiopeia	Shore slater (Ligia)

These are just a few of them, for there are endless varieties. However, the aquarium should not be overloaded with them or they will wreak havoc with the plants and quickly pollute the small water area. Go slowly with specimens. Add only a few small ones at first and very gradually add more if your jungle grows and can sustain them. A few of the most common and desirable specimens will be discussed below.

Fiddler Crabs

Fiddler crabs are one of the most common and easily obtainable denizens of the mangrove jungle. They are a fascinating little crab and mature at just over an inch or two long, making them ideal for the close confines of a home aquarium. The male fiddler has a huge claw or pincer, nearly as long as his own body. He uses this for pro-

The male fiddler crab has a huge claw. He will raise it over his head to signal his mate.

tection as well as to obtain food. But during the mating season, he uses his huge claw to signal his mate. This is what gives the fiddler crab his name, for as he raises and lowers the large claw in a sharp, jerky motion, it somewhat resembles a musician playing the fiddle. Only the male has the large claw, so it is simple to distinguish between the sexes.

Fiddler crabs will add a great deal of life to your jungle. They will soon feel right at home and you will be able to watch them "fiddle" with their large claws from time to time. It is an amusing sight. Do not overload your aquarium with them, for this will only result in fighting and the eventual ruination of your mangrove trees. Three or four fiddlers are fine for a twenty-gallon tank, eight to twelve for a fifty-gallon aquarium. All crabs feed on detritus, decayed vegetation, or small chunks of shrimp or fish. You may feed them by placing a small chunk of fine, chopped fish or shrimp mixed with a small amount of aquarium dry food in the aquarium. Place the food on the land inside a flat shell. If it isn't eaten, the shell should be removed from the tank and well washed after one day so that it won't cause any objectionable odor. The crabs will find little bits of food in the pond as well as around the mangrove plants. A few blades of grass with the root section may be placed half in and half out of the pond and allowed to rot. They will often eat this, but if it becomes too unsightly, simply remove it. Crabs are true scavengers and will eat just about anything from fruits to nuts. Feeding them will present no problem.

Hermit Crabs

There are two distinct types of hermit crabs: one that lives on land and one that lives in the water. Both will be discussed here. The land hermit, as its name implies, lives entirely on land and will actually drown if placed in deep water where it cannot come to the surface to breathe. It lives in an empty seashell that it carries about wherever it goes, and this is its home. It is a very hardy crab, eats practically any-

Small hermit crabs are excellent in the mangrove swamp or in the aquarium. This little fellow was snow white, both in body and shell. It was collected on snow-white sand and uses its white color to camouflage it as protection against sea birds. Hermit crabs found on land are different from water hermits and will drown in the aquarium if they cannot climb out to breathe.

thing, and can even live completely away from the ocean. In fact I have several living in my yard in Miami that have been there for several years and are still doing well. They eat lawn grass or whatever they can find in the yard. Our house is about thirty miles from the ocean.

Land hermits should be obtained in the small size of a half inch to an inch for your jungle. Larger specimens have very powerful claws and will quickly uproot and destroy your mangrove trees. Get the smaller specimens and put them up on the land portion of your jungle. They will fend for themselves after that. You can feed them a small chunk of apple, carrot, banana, grass, or just about anything. Feed them lightly and if they don't eat the food within a day or two, remove it from the tank so that it won't create a bad odor. They may take food right from your fingers when they have become accustomed to you.

Land hermits grow very slowly, but if you want them to be happy, you should put in an extra seashell or two so they can change houses every now and then. Everyone likes to move into a new house and hermit crabs are no exception. Give them a shell a little larger than their present home and perhaps you will be able to see them move into the new house. It is quite an undertaking. They will examine the new home from all angles, often tapping on the roof or peering inside to examine the plumbing. Then if they like the new place, they will move it into position right next to their own shell and crawl completely out of their old shell and move into the new abode. Sometimes hermits will fight over a shell, so if this occurs, be certain to put in several shells so there will be new houses for all of them. Use the gastropod-type shells with the narrow opening if possible, since they prefer these.

Hermit crabs are interesting little fellows and they will add hours of enjoyment to your mangrove jungle. As with the fiddlers, do not overload your tank with them. Three or four is plenty for the average aquarium.

The water hermit crab is similar to the land species except that it lives entirely in the water. These may be placed in the pond area of your jungle, and they will not only enliven the water area but will help to keep it clean by eating small debris that settles to the bottom. They may be fed a pea-sized chunk of fish or shrimp once a week. Push it directly beneath them if you like and they will usually grab it quickly. Two or three small half-inch to one-inch specimens are fine for the average aquarium. Like the land hermits, they will appreciate an occasional new house and will usually take greater interest in any new shell placed in their aquarium. Water hermits and any other marine life placed in the little pond will require a certain amount of oxygen in their water. Therefore, you must aerate the water in the little pond twice daily by dipping up a glass or two of water and pouring it back into the pond from a height of a foot or more so that it makes bubbles in the water. If you have other aquariums, an air stone from the air pump will help a great deal, but it is not usually needed unless you keep a great many specimens in the pond. However, an inexpensive air pump, attached to the sub-sand filter, will keep the water in better condition.

Other Crabs

There are numerous other crabs that live in and around the mangrove jungles. These include the blue crab, stone crab, mangrove crab, and porcellan crab, just to mention a few. All will live in your jungle but again, obtain them only in small size. The blue crab in

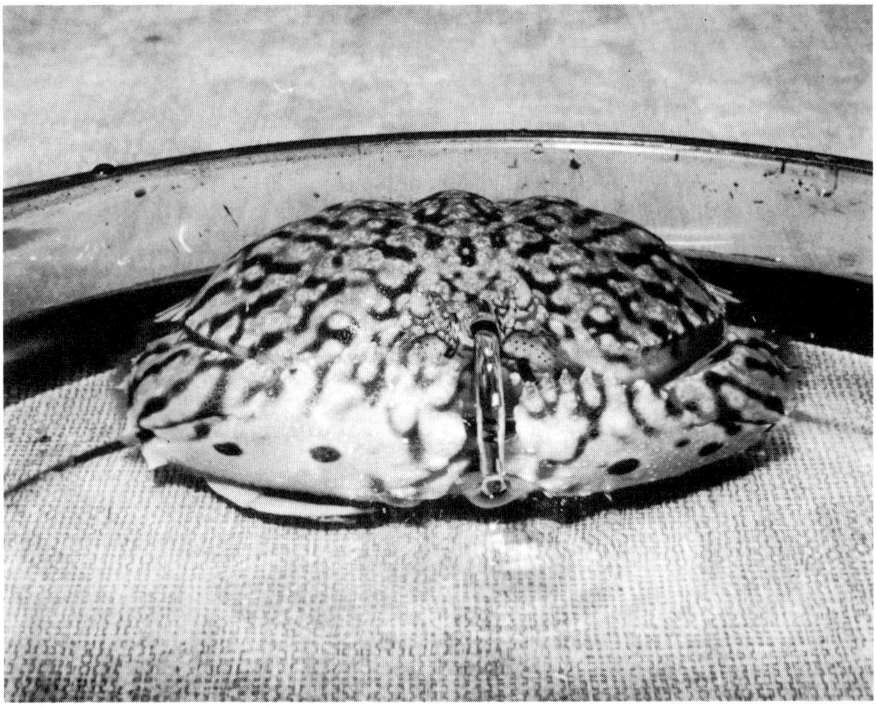

The box crab or fountain crab is often found in the shallows and is interesting in the shallow mangrove tank, for it will pump water like a fountain as shown here. I have seen them pump water six inches high. I call them fountain crabs. If they chew on the mangroves, place a rock so they cannot reach the roots.

particular is extremely vicious and will quickly attack anything else in the aquarium. Small half-inch to inch-long specimens are best. Larger specimens will simply destroy the exhibit as well as the other inhabitants. Of course if you have a very large mangrove jungle eight to ten feet across, then you could keep full-sized specimens. You might become so fascinated with the project that you would turn a spare room into a jungle, complete with pond. There are huge blue land crabs that would be most unusual in a larger exhibit.

Miscellaneous "Critters"

There are many other creatures of the shallow tidal areas that would be interesting in your jungle. If you travel to the seashore, you can collect many of them yourself or you may have them shipped to you from the ocean areas by marine collectors. The above specimens are relatively hardy and easy to obtain, but if you want more of a variety you can vastly increase your selection of inhabitants by installing a sub-sand filter under the sand when you first set up the aquarium and then hook this to an air pump to give the aquarium good aeration. Put the bubbler stem *at the front* of the tank, rather than in the back as you would a regular aquarium, so that the water will bubble back into the pond. Also, cut off the bubbler stem so it is only three inches high before placing the filter in the aquarium. The filter should be covered with one inch of sand in the front part of the tank so that the filter will operate. The back portion of course should be four or five inches deep so the trees can be properly rooted.

With this set up, the water will stay much cleaner and you will be able to add a little more marine life to your jungle, depending on the size of your aquarium and the depth and scope of your pond. Unusual critters like pygmy sea horses, pipe fish, nudibranchs, baby snook, baby tarpon, saltwater mollies, gambusia, and baby lobsters could all conceivably be kept in the jungle pond at the front of the tank—Not all of them at once of course, except in a very large tank, but a few of them certainly could be kept with good success. You cannot have as large a variety as you could with a regular saltwater aquarium because you are dealing with a small quantity of water with a mangrove jungle, for the pond is really a secondary part of the project, but if you set up a fifty-gallon or larger display, the pond at the front will then be of considerable size and a number of little specimens could then be introduced. I have written a larger, comprehensive book on saltwater aquariums that deals with the keeping, feeding, and maintenance of practically every type of marine organism that could be kept in an aquarium, which covers too broad a field to be included in this book. The large book covers it in detail.

An aquarium with giant plume worms, orange coral, and small, colorful fish is only the beginning, for nearly everything that lives in the sea can be kept alive indoors when we have acquired the knowledge.

There is really no limit to the size to which mangrove jungle can be built. Museums and public aquariums could easily construct a twenty- to fifty-foot jungle of *live* mangrove trees and a pond of such magnitude that it could harbor fair-sized snook, tarpon, and perhaps even the rare saltwater crocodile. Even small shore birds could be displayed with a real mangrove jungle (minus the croc of course). The large, blue land crabs could easily be included in such an exhibit as could raccoons, lobsters, and any number of dwellers of the mangrove swamps. These of course could only be kept with full-sized mangrove trees, not newly sprouted beans. But in this case, the trees would have to be obtained half grown and carefully transported to the exhibit. Mangrove trees are extremely hardy. I have even planted them near my house in Miami, over thirty miles from the sea.

9
The Sea Is the Limit

THERE ARE OVER 800,000 SPECIES OF INVERTEBRATES, AND THE GREAT majority of them live in the sea. There are thousands of species of fish. Many are yet to be discovered. This is what makes the saltwater aquarium, especially the live-coral aquarium, so fascinating. It is the aquarium of discovery and you are the scientist! You can discover things in your live aquarium that are not in any books and that are not even known!

We are just beginning to learn about the coral reef. We know practically nothing about it. The saltwater aquarium with a piece of the reef in it becomes a miniature laboratory, a window to the bottom of the sea, where the day-by-day and hour-by-hour processes of life on the reef can be studied in detail at first hand. It is an opportunity to learn, never before offered in our lifetime. Naturally you cannot expect to set up a coral reef in your aquarium and have the whole thing live in perfect health. What we are trying to achieve is to keep a small segment of it, a mere portion of it, alive for at least a reasonable length of time. We haven't quite progressed to the point that we can keep a whole reef alive in the aquarium, but I hope we will make progress toward this goal in the years ahead.

You will have some losses with your corals, plants, and fish. This is only natural. Everything requires experience. You could go out and buy a package of flower or vegetable seeds and plant it exactly according to the instructions. Your crop may be successful or it may be a total failure. Such things as nematodes, poor soil, insects, temperature, and rainfall all enter into the picture, as does the so-called green thumb. Growing live things just isn't that cut and dried. The same applies to your live-coral aquarium. It may be completely successful, medium, or poor. It may be a total failure. Let's hope it isn't. Medium success would be nice, and if you are fortunate, total success may be yours.

But don't give up. Even if your corals or plants partly die, they may recover and put out new growth once the aquarium becomes established and stabilizes itself. Some plants like the caulerpa may lose their fronds at first but will eventually fill up the entire tank with greenery. You may see tiny starfish emerge from your live rocks or new corals form. Tiny snails may hatch out and fill your aquarium. Tube worms may grow on once-barren rock, and sea squirts and sponges may suddenly appear. If your corals and live rocks are from Pacific and Indo-Pacific reefs, many strange, unfamiliar forms may appear in time.

One thing is for certain. With the sterile tank of bleached corals, few new things are likely to ever appear, for the corals are dead and the only things alive in the aquarium are the fish and invertebrates that you place there. But with the live coral tank, the sea is the limit. Almost any thing is likely to hatch out. I have had baby filefish, baby cowfish, squids, and octopus hatch out in a natural reef aquarium as well as dozens of different species of shells. The live rocks and corals contain the "seeds" of many different forms of marine life, and if you keep your aquarium in good condition and don't let the water become polluted from overfeeding and don't fail to remove a dead or dying coral or gorgonian, you will be rewarded with new forms of life that will start as babies and grow to maturity in your aquarium.

Deep Sea Specimens

Probably the most exciting venture in a live-rock or live-coral aquarium will be the keeping of live rocks and corals from the great depths of the sea. Although at present this unique opportunity is only available to a few scientists working on research ships with deep-water trawls, a few deep-water specimens are available at the present time from Woods Hole Marine Biological Laboratory at Woods Hole, Massachusetts. They are fairly expensive, but considering the work involved in bringing them up from the great depths alive and in good condition, the price is a bargain.

But this is just a scratch on the surface, for each year more deep-sea trawlers appear on the scene, mostly as commerical fishing boats. Nevertheless they do bring up astounding catches of deep-sea corals, live rocks, and invertebrates. The fishermen themselves are usually interested in a single product, shrimp, bottom fish, or scallops, and they generally have no time to bother with all of the fantastic, interesting forms of life. These are quickly shoveled overboard when the catch is sorted. But an enterprising marine aquarist with a business sense could go out regularly on these boats and collect all kinds of interesting live specimens, many of them new to science. It would require considerable experience in handling live specimens and there would be problems to be worked out, such as areation of the specimens in the holding buckets, perhaps a chilling of the water, changing of the water, and other factors. If the specimens are caught at great depths, say a thousand to several thousand feet, the water down there is near freezing and the specimens will need to be kept cool. This could be accomplished by ice packed around the plastic bags or by using double buckets, placing the ice between them. The problems are complex but not really difficult. The ice may not be needed at all, but if it is, usually it will be available right on the boat. It would be best to offer to pay the captain to allow you to go aboard, because you and your buckets are certain to be in the way, so that a small payment will help to offset any inconvenience you may cause them.

Once you have mastered the problems and learned which specimens are worthwhile, you could probably build up a huge business in deep-sea specimens, shipping them all over the country. Even dead specimens would be worth at least something, much more than a similar specimen from the shallow depths. But it will be the live specimens that will make your venture worthwhile, for you will be contributing to our knowledge of the sea.

Marine biology students working on the scientific research ships that trawl at the very bottom of the sea to depths of three to five miles could also make a profitable income collecting live specimens, which are in excess and not needed for the general collections of the ship. This would have to be authorized by the school or university using the ship, but perhaps if the venture could bring in income, a portion of this could be returned to the university to finance future deep-sea exploration. Some marine biology students could make a good and permanent living just collecting live deep-sea specimens, and since full-time employment in the field is often difficult to obtain, this could open up an exciting career for them.

The great depths of the sea contain a wealth of life still largely unknown to modern science. Strange, abyssal sea cucumbers have been brought up from depths of over six miles. Living fossils, specimens thought to be extinct for millions of years, have been brought up alive from the great depths. The deep-sea lily (*Rhizocrinus lofotensis*), a primitive stalked crinoid, thought to be extinct for over 160 million years, was brought up from the sea floor alive! Deep-sea urchins known only from fossil beds of millions of years ago are now brought up alive with good regularity. Rare seashells of long-extinct species now come up alive in the deep-sea trawl. Imagine keeping these strange creatures alive in the aquarium. How exciting! What else will we find in the great depths? Will the ancient dinosaur be found there in some hidden trench or cave, a real sea dragon with fire shining from its eyes? It sounds impossible, but more than a few scientists now believe that there are real sea monsters living in the depths of the sea. While it is unlikely that there would be an air-breathing dinosaur in the depths since it would be quite difficult for a beast to go all the way to the surface for an occasional gulp of air, what's wrong with a *gill* breathing dinosaur, like the newt or salamander? Couldn't such a creature exist in the great depths?

Deep-sea submarines have already reported strange giant fishes with huge armorlike scales and reptilian appearance. Deep-sea sonar has picked up creatures moving in the depths that were not whales because they never came to the surface. The occasional sighting of strange creatures or sea monsters even in our modern times gives credence to the possibility that at least some of these creatures certainly do exist.

We have just begun to explore the great depths with deep-diving subs and much more will follow in future years. Perhaps someday the deep-diving subs will be collecting live specimens as a means to generate income. One sub from Japan is reported to be harvesting precious coral. It is not unlikely that if a market can be found for certain live specimens from the great depths, then at least some of the deep-diving submarines will look into the possibility of obtaining the specimens. The larger public aquariums, always looking for something new and exciting, would certainly be interested in some live deep-sea angler fish or other deep-sea specimens. The live corals or live rocks bristling with crinoids and mollusks would certainly interest any public or private aquarium. The aquarium hobbyist of course would also be interested. A deep-sea specimen of living rock or coral could easily bring fifty dollars to one hundred dollars or more so that collecting it could be well worthwhile.

How to collect it from a deep-diving sub would be trial and error at first. I don't know what would be the best way to do it. It would appear that a half-dozen steel cylinders attached to the sub could be especially constructed to hold live specimens. These would be in an upright position near the observation windows of the sub and in reach of the sub's mechanical arm, which would collect the specimens. The specimens of live coral or rock are placed in the cylinder. Then the lid is closed. Tiny holes in the top of the cylinder, or near the top, allow some of the water to escape as the sub slowly makes its way to the surface. The specimens are protected from light, because the cylinder is completely closed. Once on the surface, the containers are removed and placed in a refrigerated room where they are kept under aeration or in very subdued light for the trip to shore. A few hundred specimens at fifty dollars each would total up to ten thousand dollars, which might even make the deep dive profitable, not counting the photographic and observation rewards.

An article in *National Geographic* (January 1971) shows photographs of the sea floor taken from the deep-diving sub *Deep Star,* at a depth of over a thousand feet. The sea floor is crawling with invertebrates of all types. The rocks are alive with arrow crabs and crinoids. What a fantastic sight one of these rocks would make in the aquarium. I believe they would live for at least a little while, a month or two, but perhaps they could live for a long while, maybe even for years. Dr. F. G. Walton Smith of University of Miami also feels they would live. The specimens from one thousand- to two thousand-foot depth might be the best because they are accustomed to at least a trace of light, so viewing them under a very dim light would probably do them no harm. Nothing of course is known about keeping or feeding live specimens from the great depths, but I suspect that they will respond to feeding much the same as the shallow water specimens, except that they may not require nearly as much food to survive. The aquarium will have to be refrigerated in all likelihood and well filtered to keep suspended matter down, for the water is both ice cold and crystal clear in the depths.

Some deep-sea specimens already have been kept alive in the aquariums and they lived for varying lengths of time, from hours to days, depending on the species. But no real attempt was made to keep them. The water wasn't even refrigerated or filtered.

The tiny squid spirula, familiar to shellers, who often find its coiled, ramshornlike shell cast up on shore, lives in the great depths. Occasional live specimens are brought up by deep-sea trawlers and some of these have been kept alive. One scientist reported that a deep-sea urchin brought up from a depth of several miles made it all the way to the surface alive.

Gorgonians, anemones, crinoids, brittle stars, sea pens, and corals are all found at great depths of the sea. Think of how exciting it would be to have a clump of living rock from a depth of three miles. Who knows what would crawl forth from it once it was settled in the aquarium? The whole thing might shine like a light. Many of the deep-sea specimens are entirely luminous, such as the sea pens. They will glow in the dark, bright enough so you can see your watch! The deep-sea fishes are luminous as well. One of the angler fishes carries a lighted lure inside of its mouth. When it wants to eat, it simply opens its huge mouth, flashes the light, and any fish that goes after it is swallowed up in a hurry.

Yes, the deep sea is still the sea of discovery. We may still find living sea monsters there, for we have not really looked at the sea bottom, except on rare occasions. It is very unlikely for a deep-sea trawl to ever catch a sea monster, for the creature could see or feel the net coming long before it arrived. One scientist made the comparison of deep-sea dredging for life samples to that of a flying machine from another planet trying to sample the life on earth by lowering a net at night and dragging it across the ground. What would he really catch? Surely not an elephant, a lion, or even a bird. He might catch some grasshoppers or a few insects and that would be all. The net would hit trees or boulders, frightening away any larger specimens and probably would get hung up and lost. The very same thing happens far under the sea.

The deep-sea trawl makes so much noise and commotion as it is dragged along the sea floor that it is not likely to catch any large, alert specimens. It hits rocks and boulders and very often the entire net is lost. It is not likely to ever catch a sea monster, but it could bring up an egg case of some unknown creature. This is a real possibility. You could very well have such a find in a cluster of deep-sea corals, should you get a hold of some live specimens. Perhaps a baby angler fish will be yours or a living fossil never before seen by man. This is what makes the deep-sea corals so fascinating, and I hope that interested aquarists can somehow get their specimens.

You could write to the universities that do deep-sea trawling, such as the University of Miami, Marine Laboratory, Woods Hole Oceanographic Institute, Scripps Oceanographic Institute at San Diego, California, and others. Or write to the companies that make the deep-diving submarines, such as Westinghouse, and tell them that you are interested in live, deep-sea corals. Or you can collect your own specimens if you have a good seaworthy boat. You will need a powerful winch and a mile of cable. Some deep-sea fishing rigs have this much and are operated electrically. Place several large grappling hooks on the end of the cable, then weight it down with lead and let it sink to the bottom. Pull it along until it snags on some rock or coral. Eventually you may snag on to a piece of ocean bottom and pull it loose. If you are working in a thousand to two thousand feet of water, every chunk of rock or coral that you bring up should be of great interest. The corals will be of the solitary type, since no coral reefs are found at that depth, but they will be interesting. Have some iced buckets of water ready for the specimens and keep them out of the bright light. You could also try a deep-water dredge. This would be made of heavy angle iron with the bag made of steel-mesh wire. It is dragged across the sea floor for the specimens. Commercial fishing houses may have them for deep-sea scallop dredging but they are certain to be expensive.

You could also make some live fish traps, baiting them with cut bait and dropping them to the very bottom. If you can leave them on the bottom for a short while, you will most likely catch some fish and invertebrates, and if they come from the deep water they are bound to be unusual and little known. But you could also work the shallow depths of three hundred to five hundred feet. This is far beyond the realm of the scuba diver, and whatever you bring up, whether it be fish, coral, or live rock, it is certain to be of interest. I once snagged a chunk of coral at six hundred feet. Another time I snagged a sponge that was full of albino coral shrimp with pink eyes! Still another time I brought up a beautiful, bright-green algae, looking almost exactly like the fresh-water aquarium plant myriophyllum.

One thing is for certain, the great depths of the sea are still an unknown world. It is not just a flat plain of deep, mudlike ooze as envisioned at one time. It is another world with mountains, deep valleys, rugged boulders, huge caves, strange forests of trees that glow in the dark, and multitudes of fish both large and small. There are flat, muddy areas of course, but there are also areas rich in life in bizarre forms, which one might expect to find on some distant planet. But it is right here, waiting for man to find it. Perhaps waiting for the aquarist, too!

Suggested Reading

Abbott, R. *Sea Shells of North America.* New York: Golden Press, 1968.

Bennett, Isobel. *The Fringe of the Sea.* Sydney, Australia: Rigby Ltd., 1966.

Maurice Burton and Robert Burton. *The International Wildlife Encyclopedia.* New York: Marshall Cavendish Corp.

Friese, U. *Marine Invertebrates.* Neptune, N.J.: TFH Publications, 1973.

Herring, P. *Deep Oceans.* New York: Praeger Publishers, 1971.

Idyll, C. P. *Abyss.* New York: Thomas Y. Crowell Co., 1964.

Miner, R. W. *Field Book of Sea Shore Life.* New York: Putnam, 1950.

Carleton Ray and Elgin Ciampe. *The Underwater Guide to Marine Life.* South Brunswick and New York: A. S. Barnes and Co., 1956.

Salt Water Aquarium Magazine. P. O. Box 1000, Belleview, Florida, 32620.

Straughan, Robert P. L. *Exploring the Reef.* South Brunswick and New York: A. S. Barnes and Co., 1968.

———. *The Marine Collector's Guide.* South Brunswick and New York: A. S. Barnes and Co., 1973.

———. *The Salt Water Aquarium in the Home.* South Brunswick and New York: A. S. Barnes and Co., 1968.

Waters. *Salt Water Aquariums.* New York: Holiday House, 1967.

Index